Survey on Advanced Fuels for Advanced Engi

Survey on Advanced Fuels
for Advanced Engines

Project report
Funding by IEA Bioenergy Task 39

tac
technologietransfer automotive
hochschule coburg

Norbert Grope
Olaf Schröder
Jürgen Krahl

DBFZ

Franziska Müller-Langer
Jörg Schröder
Eric Mattheß

Date: October 2018

IEA Bioenergy

Bibliografische Information der Deutschen Nationalbibliothek

Die Deutsche Nationalbibliothek verzeichnet diese Publikation in der
Deutschen Nationalbibliografie; detaillierte bibliographische Daten sind im Internet
über http://dnb.d-nb.de abrufbar.

1. Aufl. - Göttingen: Cuvillier, 2019

© CUVILLIER VERLAG, Göttingen 2019
  Nonnenstieg 8, 37075 Göttingen
  Telefon: 0551-54724-0
  Telefax: 0551-54724-21
  www.cuvillier.de

1. Auflage, 2019
Gedruckt auf umweltfreundlichem, säurefreiem Papier aus nachhaltiger Forstwirtschaft.

  ISBN 978-3-7369-7036-6
  eISBN 978-3-7369-6036-7

Contents

Abbreviations

AFID ....Alternative Fuels Infrastructure Directive
AFV ....Alternative Fuel Vehicles
AMF .... Advanced Motor Fuels
BEV .... Battery Electric Vehicle
BtL... Biomass to Liquid
BTX ....Benzene, toluene and xylene isomers
CFPP.. Cold Filter Plugging Point
CI.. Compression Ignition
CN .... Cetane Number
CNG...Compressed Natural Gas
CP...Cloud Point
CPD ....Clean Power for Transport Directive
d/iLUC..direct/indirect Land Use Change
DME .... Dimethyl ether
DOC.. Diesel Oxidation Catalyst
DPF ....Diesel Particle Filter
EGR .... Exhaust Gas Recirculation
EPA .... Environmental Protection Agency
ETBE ....Ethyl Tert-Butyl Ether
ETD ....Energy Taxation Directive
EV.. Electric Vehicle
EU .... European Union
FAME ....Fatty Acid Methyl Ester
FCV.. Fuel Cell Vehicle
FEV.. Full Electric Vehicle
FQD ....Fuel Quality Directive
FT .... Fischer-Tropsch
FT-IR..Fourier Transformed Infrared Spectroscopy
FRL .... Fuel Readiness Level
GC/MS ....Gas Chromatography/Mass Spectroscopy
GHG ....Green House Gas
GTL..Gas to Liquid
HC ....Hydrocarbon
HCCI ....Homogeneous Charge Compression Ignition
HDV... Heavy Duty Vehicle
HEFA .... Hydroprocessed Esters and Fatty Acids
HVO ....Hydrotreated Vegetable Oils
IEA.. International Energy Agency
LBG...Liquefied Bio Gas
LBM .... Liquefied Bio Methane
LCA... Life Cycle Assessment
LCFS ....Low Carbon Fuels Standard
LDV .... Light Duty Vehicle
LEV .... Low Emission Vehicle
LG..Liquid Gas
LNG ....Liquefied Natural Gas
LPG..Liquefied Petroleum Gas
MDV..Medium Duty Vehicle
NM[V]HC..Non-Methane [Volatile] Hydrocarbons
MMT .... Methyl-cyclopentadienyl-Manganese-Tricarbonyl
MTBE ....Methyl-Tert-Butylether

Executive summary

The literature study "Survey on Advanced Fuels for Advanced Engines" has been set up as a review-like compilation and consolidation of relevant information concerning recent and upcoming advanced engine fuels for road vehicles with special focus on biomass-based liquid fuels. It is provided as a self-contained report, but at the same time serves as an updated and complementary resource to IEA-AMF's online fuel information portal (http://www.iea-amf.org). An attempt is made to describe the *status quo* and perspectives of advanced fuels and to give a broad overview on parameters, tools and experimental approaches necessary for fuel characterization and evaluation. The focus of literature coverage, especially concerning fuel properties and exhaust emission research results, is from recent to approximately five or ten years back, but if appropriate, older resources were considered too in the general discussion of relevant effects and mechanisms.

Introductory Chapter 2 summarizes framework conditions for advanced fuel applications in terms of regulatory measures and incentives for sustainable and fair-trade action, climate change prevention and energy-efficient vehicle operation. Following these non-technical topics, Chapters 3 and 4 gives information about fuel standards and fuel properties, which should be considered when introducing new fuels. Chapter 5 provides tabulated information on feedstock, production schemes, costs and market issues for the main types of advanced biofuels considered in this study, i.e. hydrotreated vegetable oils (HVO/HEFA), biomass-to-liquid (BTL) fuels (i.e. paraffinic Fischer-Tropsch (FT)), methanol, dimethyl ether (DME), oxymethylene dimethyl ether (OME), lignocellulosic ethanol and liquefied biomethane. Also fuel properties and emission trends are shown in this chapter. Accordingly, biodiesel is explicitly included in subsequent discussions and complemented by an excursus on metathesis biodiesel.

Chapter 6 refers to reactivity and stability of fuels with regard to interactions among different fuel components and between fuel and engine oil. Deterioration of fuel and engine oil quality will affect long-term fuel storage and vehicle functionality by formation of deposits and sludge and is manifested by laboratory parameters not necessarily detectable macroscopically. Influencing factors like molecular structure, temperature, oxidizing agents, additives, impurities and metal catalysis are discussed according to published research results.

Chapter 7 deals with health effects of engine exhaust and to this end describes important gaseous and particulate constituents, their characterization and measurement. As specific exhaust species, regulated parameters CO, HC, NO_x and unregulated components polycyclic aromatic hydrocarbons (PAH) and carbonylics are considered, and particular aspects of particle size, number and composition are discussed. Reference is made to formation of ozone and ambient aerosol as secondary impacts of engine exhaust. Short keynotes on research and review articles on issues of toxicology, mutagenicity and other adverse effects of engine exhaust are provided. A thorough introduction to dedicated engine emission testing and a literature survey on published emission measurement results using various engine types and fuels are given.

As a conclusion of the study, the diversity of fuels will increase in the future. New advanced fuels will be introduced in the market (e.g. HVO) or will become the focus of research activities (e.g. OME). One criterion for successful introduction of a new fuel in the market is that the new fuel can be used as a drop-in fuel. These fuels have the advantage that small amount of the fuel can be tested using existing infrastructure and engine techniques. In this phase of market introduction, reactions among fuel components and material interactions can also be detected. At the moment, most research activity deals with the behavior of aging products of biodiesel in non-polar fuels like HVO/HEFA or X-to-liquid (XTL, FT fuels).

Introducing new fuels needing an adaption of the engine technique or a new engine concept in the market requires much more effort. Next to the new fuel, also a new infrastructure and new engines have to be developed and launched. This is only possible if fuel and automotive industries, politics and broad public support the new development.

Another key factor for advanced fuel is the raw material base. The production of advanced fuels should be independent of fossil resources. Therefore, biomass or renewable electrical power (e.g. wind power or solar energy) must be the source of advanced fuels. Biomass is intensively used by first generation biofuels, but there is a potential to raise the share of renewable fuel with the introduction of advanced fuels having a broader base of biomass. Electric power as an energy source for advanced fuels also will become interesting, if the share of renewable electricity in the grid will increase. Nevertheless, already today research is necessary to have the technique(s) ready in time.

Last but not least, for further development of engine technique, advanced fuels can be use as construction or design element. If it is possible to optimize the burning process and to minimize emissions by the use of advanced fuels, new vehicles can have a better performance at the same price.

In summary, advantages and disadvantages of the considered advanced fuels are listed in Table 1. From today's point of view, no advanced fuel has the potential to fully replace fossil fuel use in the near or middle future, but all advanced fuel options have the potential reduce fossil fuels usage significantly.

Table 1: Advantages and disadvantages for the market introduction of advanced fuels (++ clear positive impact, + slight positive impact, 0 no impact, - slight negative impact, -- clear negative impact)

Fuel	Production technique	Raw material base	Drop-In fuel	Engine technique	Exhaust gas emissions
HVO	++	+	++	++	+
BTL	0	+/++	++	++	+
DME	0/++	methanol	--	0	++
OME	--	methanol	0	0	++
Methanol	0/++	+/++	-	0	0
Lignocellulosic ethanol	+	+/++	+	+	+
Bio-LNG/LBM	++	+	++	++	0 (++)*

*compared to fossil methane (to gasoline)

1 Introduction

1.1 Background and motivation

The long history of technical evolution of combustion engines and appropriate fuels has seen many adaptations and refinements that underline the close interrelation of engine and fuel development. Operation of engines by applying "whatever may serve as combustive fuel" is an outdated concept, since modern engines are highly sophisticated instruments requiring clean, well-specified operating fluids as fuels. The fuel requirement mutates from the former simple energy carrier to a future key constructional element for combustion engines.

Today's road transport still almost completely relies on vehicles powered by combustion engines, with fossil fuels being the main energy carrier. The sheer dimension and ongoing increase of road traffic leads to large amounts of fossil fuels being burnt, coupled to corresponding pollutant emissions and consumption of resources and energy for fuel supply. The urgency of reversing negative trends of climate change and environmental pollution demands countermeasures that on one hand reduce fuel consumption in general and enhance engine efficiency in particular. On the other hand, a broader range of adverse emissions and unwanted effects associated with fuel supply and road traffic has to be controlled and minimized. To this end, it is necessary to improve engine design as well as to optimize fuel properties in terms of engine performance, sustainability and climate and environmental preservation.

Engineering efforts to cope with these technical challenges are embedded in general social and market requirements like security of energy supply, diversification of fuel sources to buffer against the instabilities of fossil fuel prices, consumer demands regarding vehicle performance, objections to vehicle concepts or compatibility of prospective alternative fuels, or objections to environmental footprints of biofuels, all under the industrial premise of cost-efficiency and competitiveness. Perception of environmental trends by the public and political implementation of corresponding regulatory measures strongly influence the margin or balance or prioritization for continuing existing technologies versus focusing on implementing new advanced technologies.

The remarkable progress achieved to date regarding engine and fuel issues has been accomplished by performing and evaluating an enormous number of engine tests and calibration experiments complemented by mathematical modeling. When referring to engine tests, we mean a single combination of combustion instrument (engine) and fuel. Results will change using a different engine or combustion apparatus, or by modifying driving or ambient conditions, and of course using a different fuel. It is therefore essential to refer to distinct engine-fuel combinations and to develop fuels and modern engines in a strongly coordinated manner. Specific engine performance (for a given fuel) implies specific exhaust gas components and consequently dictates the requirements for the exhaust gas aftertreatment strategy. Moreover, fuel chemistry is important for fuel-fuel and fuel-engine oil interactions, especially to functional chemical groups potentially present in the fuel that increase or decrease polarity or are susceptible to oxidation, like e.g. unsaturated carbon bonds. Undesired effects are formation of deposits in the fuel storage and supply system and sludge in the engine oil that can severely damage engines. Whether in neat form, as preformulated blend or as drop-in application, limited suitability of fuels for long-term storage in the fuel tank is a critical factor for plug-in hybrid electric vehicles (PHEV) that have low/infrequent fuel consumption and prolonged intervals between refillings. Emergency stand-by generators, seasonal vehicles and machinery with only sporadic use share these same issues and raise similar concerns.

A suite of advanced fuels including biofuels have emerged over the years whose production continues to be optimized with respect to sustainability, greenhouse gas mitigation, engine performance and exhaust gas quality. Broad acceptance of a new advanced fuel, meanwhile, depends on proper experimental and practical experience from engine/vehicle testing as well as reliable and sufficient supplies of such fuels to enable engine tests to be performed. Awareness of corresponding fuel properties both to experts and consumers is also a prerequisite for general acceptance, market entry and for focused research and development (R&D) to further optimize engine combustion.

Scientific knowledge on chemical species and inventories of vehicle pollutant emissions, underlying mechanisms and influencing factors has grown substantially over time, as has knowledge on pollutant effects on biota and ecosystems. In parallel, powerful techniques have been developed for detecting and monitoring relevant processes and chemical species over much of the lifecycle of fuels usage (transport and storage, combustion, exhaust treatment). Insights into factors influencing fuel chemistry, fuel stability and combustion behavior have helped to establish concepts for tailor-made fuels and quality criteria with respect to substance class composition and purity.

1.2 Objectives and structure of the study

To contribute to positive developments for the future use of especially advanced biofuels, a global survey on different types of biomass-derived fuels and their qualities is necessary; similar to what is provided by OEMs for fossil fuels. This knowledge is relevant to identify potential future problems, challenges and opportunities for state of the art and advanced engine technologies – independent of whether biofuels are used as oxygenates or blends, or as neat or drop-in fuels. A dedicated survey of the global situation on current and advanced biofuels options enables OEM and Tier1 suppliers to optimize their modern and advanced engine technologies in accordance with regulatory requirements.

Past experiences and lessons learned suggest that such a joint implementation of engines and fuels development might have been a mechanism for reducing concerns, obstacles and constraints from the automotive industry against the use of biofuels. With regard to the targets and objectives of IEA Bioenergy Task 39, this study contributes a survey on dedicated quality aspects of certain liquid biofuels with special regard to advanced biofuels. Due to the fact that biofuels face special sustainability concerns in terms of exacerbating possible fuel-food conflicts, land-use change and engine performance problems, a sound and valid data basis is essential to facilitate the discussion. To attain a comprehensive characterization of advanced fuels, these are considered from multiple perspectives: (i) with respect to fuel properties within the regulatory frame of fuel standards; (ii) by looking at possible unintended fuel reactions, interferences and resulting vehicle performance issues; and (iii) regarding health and other effects of engine emissions as well as factors that impact exhaust quality and proper measurement of emission data.

2   Non-technical framework for advanced biofuels

Impacts of road transport on global climate, environment and human health have become an increasingly important issue of public debate and political decision making where substantial technological progress is needed. It is well recognized that straightforward sustainability criteria such as greenhouse gas reduction, protection of valuable ecosystems and avoidance of food-fuel conflicts and promotion of fair-trade in general, are important goals to be achieved. Consequently, mandatory regulations and incentive measures have been implemented in order to make fuels and vehicles fulfill minimum requirements and stimulate necessary improvements with respect to criteria mentioned.

While this study focuses on modern engine fuels as the technical factor of key importance for clean and efficient engine operation, it is useful to examine the general regulatory framework concerning sustainable mobility. Therefore, we will introduce the subject of advanced fuels by providing a tabulated, summary overview of recent and near future (anticipated) policies, regulations and incentives for fuel supply and transport. This covers

- Political and legislative aspects on an international level;
- Current development of biofuel markets and forecast scenarios for different biofuels;
- Discussion of potentials and challenges of advanced fuels;
- Requirements with regard to sustainability criteria.

Fields of action and relevant factors that characterize the non-technical framework concerning fuel and vehicle regulation are shown in Figure 1.

Regarding technical aspects, increasingly restrictive regulatory measures have been implemented to limit the adverse effects of motor vehicle traffic. Corresponding regulations specify engines and vehicles, emissions from driving operation, physical-chemical properties of fuels and operating fluids as well as requirements for performing test procedures and laboratory analyses. These issues are discussed in Chapter 4 (Technical background of selected fuel properties) and Chapter 7 (Known health effects).

The following Table 2 to Table 14 are a summary of facts on regulatory and incentive measures in relation to advanced fuels currently implemented or discussed among IEA Task 39 and IEA-AMF Annex 52 member countries as well as China as it is so important within the global market. Facts have been elaborated in various country reports by Task 39 members and in a recent publication from IEA-RETD, "Driving renewable energy for transport – Next generation policy instruments for renewable transport (RES-T NEXT)", available at http://iea-retd.org/archives/publications/res-t-next. See Figure 1 on next page for a visualization of relevant contents and fields of action.

According to key messages given in the executive summary (page 1) of this IEA-RETD document, *"the most effective instruments for increasing the share of Alternative Fuel Vehicles (AFVs) are:*

- *Zero Emission Vehicle (ZEV) mandates (obliging OEMs to meet a minimum share of ZEVs in their sales);*

- *Financial incentives in vehicle registration taxes (VRT) and in company car taxation;*

- *CO_2 regulations of road vehicles, particularly when CO_2 targets are sufficiently ambitious."*

With respect to increasing contributions of renewable energy carriers, it is claimed that fuel regulations and renewable energy mandates are the most effective tools. In order to attain the highest possible greenhouse gas emission reductions, it is recommended to combine measures promoting a higher share of renewable fuels within the fuel mix with regulations requiring improved fuel efficiency and reduced CO_2 (GHG) emissions from vehicles.

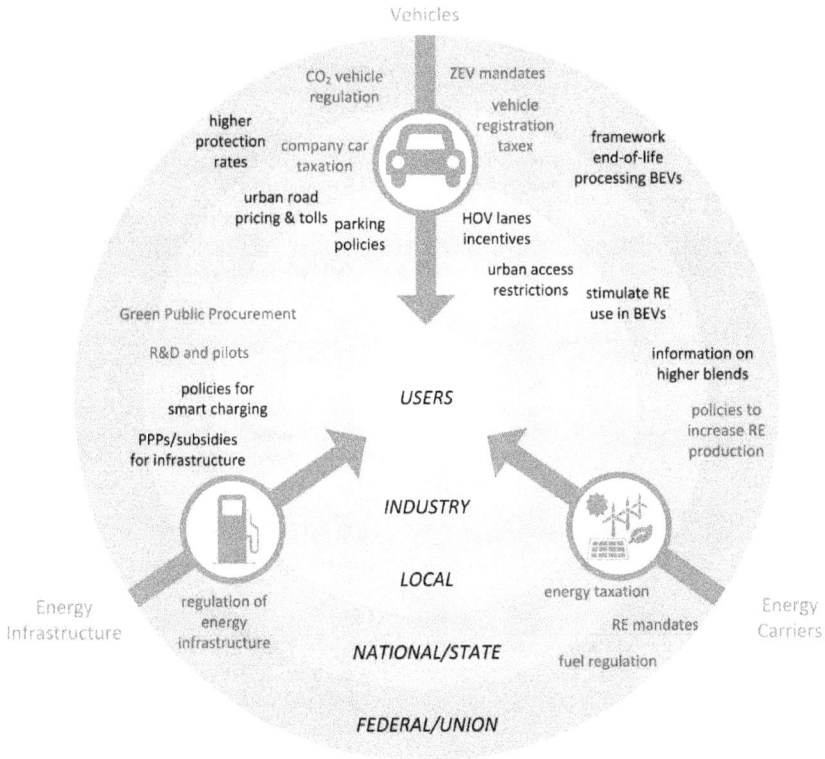

Figure 1: Fields of action and relevant factors that characterize the non-technical framework concerning fuel and vehicle regulation. Adapted from, and illustrating summary results of, IEA-RETD-RES-T-NEXT 2015, driving renewable energy for transport – next generation policies (http://iea-retd.org/archives/publications/res-t-next), compiled in Table 2 through Table 14.

Comparing the activities from different countries (Table 2 to Table 14) shows a wide range of established regulations. Mandates for alternative fuel range from zero in Norway and South Korea to 20% renewable fuel in Sweden and 27% bioethanol and 10% biodiesel in Brazil. In addition, the strategies to reach these goals differ strongly between countries. Tax releases for alternative fuels, blending mandates, reducing fleet average CO_2 emission limits, and grants for low or zero emission vehicles are the most common instruments to reduce CO_2 emissions. Worldwide, there is no uniform strategy to introduce alternative fuels in the market and reduce CO_2.

Table 2: EU – summarized from IEA-RETD report REST-T NEXT

Topic	EU
Energy carrier – RE mandates	Fixed in RED 2009/28/EG with 10% RE in 2020 and in FQD with 6% CO_2-eq reduction by 2020; Note: The REDII (Council Version of June 2018, (2016/0382 (COD)) specifies a 14% renewable energy target in transport by 2030. It is not yet officially published, but considered an 'agreed' version.
Legislation for biofuels/renewable fuels	RED: 7% cap for biofuels based on food-based crops, i/dLUC, sustainability criteria and methodology for GHG; min. 60% GHG reduction for new plants
Incentives for advanced biofuels	Determined in RED amendment, sub-targets for advanced biofuels
Taxes for fuels	Energy *Taxation* Directive (ETD) unfavorable
Energy infrastructure	Clean Power for Transport (CPD): only marginal steering in favor of advanced fuels Alternative Fuels Infrastructure Directive (AFID): setting frame conditions and facilitate installation of infrastructure, mainly electric/hydrogen/ natural gas, biofuels and reformulated fuels not a priori excluded
Vehicles – Legislation for vehicle emission standards and fuel efficiency	CO_2 regulations 2020 on tank-to-wheel basis, fleet averages for manufacturers: 95 g km^{-1} passenger cars and 120 g km^{-1} for LDV Euro 6 emission regulations HC, CO, NO_x, PM, PN. Possible inclusion of hitherto non-regulated pollutant parameters
Incentives for purchasing ZEV/LEV	See individual tables for EU member-countries (Table 7 to Table 10)

References: RED – Renewable Energy Directive: (2009/28/EC); FQD – Fuel Quality Directive: Directive (EU) 2015/1513 of the European Parliament and of the Council of 9 September 2015, amending Directive 98/70/EC relating to the quality of petrol and diesel fuels and amending Directive 2009/28/EC on the promotion of the use of energy from renewable sources; http://ec.europa.eu/clima/policies/transport/fuel/documentation_en.htm; http://ec.europa.eu/clima/policies/transport/vehicles/vans/index_en.htm; http://ec.europa.eu/clima/policies/transport/vehicles/heavy/documentation_en.htm

Topic	US
Energy carrier – RE mandates	Renewable Fuels Standard (RFS) in California: renewable volume obligation with stepwise increases until 2022, GHG reduction effect versus fossil fuel as criterion to be accepted as "renewable": -20% for "renewable fuel", -50% for "advanced biofuel".
Legislation for biofuels/renewable fuels	Low Carbon Fuels Standard (LCFS): requires 10% reduction in carbon intensity of transportation fuels by 2020 compared to 2010. CI on life-cycle basis as grams of carbon dioxide equivalent per unit energy of fuel (g CO_{2e} MJ^{-1}).
Incentives for advanced biofuels	Credit trading as implemented in LCFS
Taxes for fuels	
Energy infrastructure	Alternative and Renewable Fuel and Vehicle Technology Program (ARFVTP) in California: Funding of development and installation of infrastructure, similar programs in some other states
Vehicles – Legislation for vehicle emission standards and fuel efficiency	corporate average fuel economy standards (CAFE): fuel efficiency of new cars at least 35 miles per gallon by 2020; National Fuel Efficiency Program: determination of fleet-wide minimum fuel efficiencies and maximum allowed GHG emissions of 163 g CO_{2e} mi^{-1} by 2025
	Environmental Protection Agency (EPA) set the standards for GHG emission and fuel efficiency of heavy duty vehicles for the time period 2021 to 2027 (Phase 2) leading to 25% GHG reduction and 5% reduction in fuel consumption in comparison to the Phase 1 (2014 to 2018) goals. (EPA 2016)
Incentives for purchasing ZEV/LEV	ZEV program in California: regulation/obligation for minimum sales of ZEV/LEV vehicles rising to 15% by 2025, similar laws in 9 other US states

Reference: IEA-RETD report REST-T NEXT

Table 4: Task 39 members: Canada – summarized from IEA-RETD report REST-T NEXT

Topic	Canada
Energy carrier – RE mandates	Renewable Fuel Regulation (RFR) requires nationwide average of 5% v v^{-1} RE in (gasoline) fuels; 2% v v^{-1} renewable fuel content in diesel fuel with some exemptions, some provinces add their own mandates exceeding the national mandate
Legislation for biofuels/renewable fuels	Some provinces (British Columbia, Alberta Ontario) have or will connected their blend mandates with GHG-reduction obligation (up to 70%)
	Fuel qualities are regulated by standards (CGSB) and meet a higher quality than the US standards.
Incentives for advanced biofuels	There are no special incentives for advanced biofuels. Never mind, there are federal incentives paid for ethanol (0.03 CAD l^{-1}) and biodiesel (0.04 CAD l^{-1}), which will end in 2017.
Taxes for fuels	Ethanol import tax for selected countries of origin (e.g. Brazil): 0.05 CAD l^{-1}
	In 2017 Alberta will apply a carbon tax of 20 CAD t^{-1} (increasing to 30 CAD t^{-1} in 2018) for conventional fuels.
Energy infrastructure	
Vehicles – Legislation for vehicle emission standards and fuel efficiency	Mandatory GHG emission limits for light duty vehicle (LDV) from 2011 on, for medium duty vehicle (MDV) and heavy duty vehicle (HDV) under consideration. Proposed regulations for GHG reduction and fuel economy following US regulations, with stepwise GHG emission reduction of 5% annually until 2025.
	The latest standards are set by TIER 3 and will be in law from 2017 reducing the emissions up to 80% and limiting the Sulphur content in gasoline to 10 ppm.
Incentives for purchasing ZEV/LEV	

References: USDA GAIN Report "Canada Biofuel Annual 2016"

Table 5:   Task 39 members: Japan – summarized from IEA-RETD report REST-T NEXT

Topic	Japan
Energy carrier – RE mandates	Act on the Promotion of the Use of Non-Fossil Energy Sources and Effective Use of Fossil Energy Source Materials by Energy Suppliers from 2009: Use of biofuels is mandatory, especially of sustainable ethanol accompanied by GHG emission reductions of about 50%; stepwise increase of volumetric targets for ethanol aiming for a blend rate of 3% v v^{-1}
Legislation for biofuels/renewable fuels	The Gasoline Quality Assurance Law limits the maximum blend rate for gasoline to 10% ethanol or 22% ethyl tert-butyl ether (ETBE). The minimum GHG reduction of fuel ethanol has to be 50% compared to gasoline.
Incentives for advanced biofuels	
Taxes for fuels	The gas tax of 53.8 JPY l^{-1} is reduced by 1.6 JPY l^{-1}, if a minimum blend rate of 3% is met. In the most cases import taxes does not occur due to tax exemptions for ETBE in general and main supplier countries in special.
Energy infrastructure	Next-Generation Vehicle Strategy 2010: Mainly focusing on electric vehicles of any type but open to clean diesel development, promoted by R&D projects Strategic Road Map for Hydrogen and Fuel Cells 2014: pushing forward the hydrogen infrastructure for fuel cell vehicle (FCV)
Vehicles – Legislation for vehicle emission standards and fuel efficiency	Top Runner Program for automobiles since 1998 and amended 2015: fuel efficiency standard for passenger car is increased from 12.7 to 23.2 km l^{-1} in FY2015 to 16.9 to 28.1 km l^{-1} in FY2020 and for van and truck from 7.9-18.2 km l^{-1} to 10.2-21.0 km l^{-1}. Next-Generation Vehicle Strategy 2010: setting targets for market share of ZEV/LEV and clean diesels in 2020 and 2030 Euro 5/6 equivalent standards for LDV/HDV (UNEP 2016)
Incentives for purchasing ZEV/LEV	Subsidies are paid for purchasing EV based on the price difference to gasoline cars. A maximum subsidy of 850,000 JPY (≈7,800 USD) is possible. (IEA 2016)

References: USDA GAIN Report "Japan Biofuel Annual 2016"

Table 6: Task 39 members: China – summarized from IEA-RETD report REST-T NEXT

Topic	China
Energy carrier – RE mandates	Target set with Five-Year Bioenergy Development Plan 2011-2015: 4 million tons of ethanol production by 2015, no further nationwide obligations for biofuel use; instead E10 mandates in force at several provinces
Legislation for biofuels/renewable fuels	Methanol as fuel available in several provinces Ethanol Production and market is fully controlled by the state including fixed ethanol and corn prices (until Oct. 2016). Due to an end of the corn price regulation in October 2016 falling corn prices and a raise in the offered volume is expected. Biodiesel production is operated by small, private companies.
Incentives for advanced biofuels	Non-grain-based ethanol: 750 to 800 CNY t^{-1} (around 120 USD t^{-1}, does not count for imported ethanol)
Taxes for fuels	Grain-based ethanol: full VAT and 5% Excise Tax Non-grain-based ethanol: VAT and Excise Tax free (does not count for imported ethanol) UCO-based biodiesel: 0.8 CNY l^{-1} tax exemption
Energy infrastructure	Mostly targets and intentions for gas filling and electric charging points; in 2014 installation of the State Council on Accelerating Guidance to Promote the Application of New Energy Vehicles managing incentive measures for operation of new vehicles.
Vehicles – Legislation for vehicle emission standards and fuel efficiency	Targets for new energy vehicles incl. methanol in some provinces ranging from 10% to 40% share. National targets for alternative fuel vehicles. National Energy Conservation Law 2008: From 2012 to 2015, corporate average fuel consumption (CAFC) levels for passenger and fuel-efficient vehicles shall meet the 2015 targets of 6.9 l per 100 km and 5.9 l per 100 km, respectively, lowered to 5.0 l per 100 km and 4.5 l per 100 km, respectively, by 2020. Local introduction of Euro 6 limits (Beijing); Euro 4 is the current nationwide standard, but will be replaced by Euro 5 in 2017. Nationwide mandate for ultra-low sulfur fuels (<10 ppm S) starting in 2017 in context of Euro 5 introduction. (UNEP 2016)
Incentives for purchasing ZEV/LEV	Subsidy scheme for electric vehicles (EV), PHEV and FCV applicable in 2016 to 2020. In special acquisition and excise, tax does not occur for EV, which is leading to savings from 35,000 to 60,000 CNY (6,000 to 10,000 USD). (IEA 2016)

References: "The Potential of Biofuels in China" IEA Bioenergy Task 39

Table 7:   Task 39 members – EU countries

Topic	Austria	Denmark
Energy carrier – RE mandates	Biodiesel 6.3 cal% Bioethanol 3.4 cal%	The actual mandate for biofuel is set to 5.75% of a company's total annual fuel sale. Petrol and diesel fuel for transportation have to contain at least 1% biofuel. (RES 2016a) A 2.5% blending mandate for 2nd generation biofuels is under development
Legislation for biofuels/renewable fuels		
Incentives for advanced biofuels		Currently there is no support for 2nd generation bioethanol neither for 1st generation fuels.
Taxes for fuels	100% bio-based fuels (unblended) are mineral oil tax-free. Other tax exemptions are made by usage (e.g. "Agrardiesel"). The tax rate for fuel varies by biofuel blend rate, sulfur content and type. Petrol (E4.6; S ≤10 mg kg^{-1}): 0.482 EUR l^{-1} Diesel (B6.6; S ≤10 mg kg^{-1}): 0.397 EUR l^{-1} (RIS 2016)	Petrol (unblended): 4.243 DKK l^{-1} Petrol (E4.8): 4.170 DKK l^{-1} Diesel (unblended): 2.695 DKK l^{-1} Diesel (B6.8): 2.681 DKK l^{-1} Tax changes every year, because the tax is based on the net price index. (RES 2016b)
Energy infrastructure		
Vehicles – Legislation for vehicle emission standards and fuel efficiency		
Incentives for purchasing ZEV/LEV	Austria announced federal incentives for purchasing EVs, starting in 2017. Up to 4,000 EUR will be paid for a BEV/FCV and 1,500 EUR for PHEVs. The participation is limited by basic car price (max. 50,000 EUR), 100% RE-power supply and more. Also electric motor bikes and commercial vehicles will be promoted. Total spending will be 55 million EUR, of which 24 million EUR are paid by automotive industry. (BMVIT 2016)	

Reference for Austria: F.O.Licht's World Ethanol and Biofuels Report (Vol. 11, Issue 20, 2013)

Table 8:   Task 39 members – EU countries

Topic	Finland (formerly with T39)	Germany
Energy carrier – RE mandates	The obligate quota for the year 2016 is 10%. The quota raises each year up to 20% in 2020. (RES 2016c)	
Legislation for biofuels/renewable fuels	Biofuels produced from waste, residues, cellulose or lignocellulose counts twice for the biofuel quota. (RES 2016c)	
Incentives for advanced biofuels		
Taxes for fuels	The exact tax for fuel depends on the used fuel components. The minimum tax rates are: Petrol: 0.359 EUR l^{-1} Diesel: 0.330 EUR l^{-1} (RES 2016d)	For each fuel type occurs an Energy tax. VAT is paid on top of the Energy Tax. Energy Taxes: Diesel (S ≤10 mg kg^{-1}): 0.4704 EUR l^{-1} Petrol (S ≤10 mg kg^{-1}): 0.6545 EUR l^{-1} LPG/CNG: 0.1800 EUR kg^{-1} Tax discounts are granted for special usage (e.g. "Agrardiesel") and LPG/CNG (until 2018) For biogas does not occur an Energy tax (BMJV 2016)
Energy infrastructure		German Federal Government provides 300 million EUR for the construction of EV charging stations. (DBR 2016)
Vehicles – Legislation for vehicle emission standards and fuel efficiency		
Incentives for purchasing ZEV/LEV		Owner of Electric vehicles are exempted from motor vehicle tax for 10 years. Since July 2016 a maximum bonus of 4000 EUR is paid for EVs, FCVs and PHEVs, which have to cost less than 60.000 EUR (excl. VAT). The bonus fund is limited to 1.2 billion EUR. (DBR 2016)

Topic	Italy	The Netherlands
Energy carrier – RE mandates	Blending mandate 5 cal%	Blending mandate 7 cal%, raising each year by 0.75 cal% until 2020 (10 cal%)
Legislation for biofuels/renewable fuels		
Incentives for advanced biofuels		
Taxes for fuels	Petrol: ≈0.96 USD l^{-1} Diesel: ≈0.98 USD l^{-1}	Petrol: ≈1.00 USD l^{-1} Diesel: ≈0.60 USD l^{-1}
Energy infrastructure		
Vehicles – Legislation for vehicle emission standards and fuel efficiency		
Incentives for purchasing ZEV/LEV		Registration Tax does not apply to ZEV. All other cars are taxed by their CO_2-emission with a huge advantage for LEVs (IEA 2016)

Reference: F.O.Licht's World Ethanol and Biofuels Report (Vol. 11, Issue 20, 2013)

Table 10: Task 39 members – EU countries

Topic	Norway	Sweden
Energy carrier – RE mandates	Blending mandate 20% up to 2020. (LAN 2016)	Swedish government announced the goal of 20% share for renewable fuels in the transportation sector until 2020. Currently there are no blend mandates in force.
Legislation for biofuels/renewable fuels	The maximum blend rate for diesel was set to 7% v v^{-1} FAME. (LAN 2016)	The maximum blend rate for diesel and petrol was set to 7% v v^{-1} FAME respectively 10% v v^{-1} ethanol. (GOS 2016)
Incentives for advanced biofuels		Over 15% tax discount for HVO. (GOS 2016)
Taxes for fuels	Petrol: ≈1.00 USD l^{-1} Diesel: ≈0.77 USD l^{-1}	Taxes for fossil fuel were already increased in 2016 and will further grow based on the consumer price index until 2020. Due to an exemption from EU regulations there are tax advantages for biofuels compared to fossil fuel until 2018 (liquid fuels) and 2020 (gaseous fuels). (GOS 2016)
Energy infrastructure		
Vehicles – Legislation for vehicle emission standards and fuel efficiency		
Incentives for purchasing ZEV/LEV	EV are exempted from purchase tax. VAT exemption is only granted for battery electric vehicles. (IEA 2016)	Vehicles with a CO_2-emission of less than 50 g km^{-1} receive a 40,000 SEK (≈4,000 EUR) discount. (IEA 2016)

Topic	Australia	Brazil
Energy carrier – RE mandates	No national fuel mandate New South Wales: 6% ethanol and 2% biodiesel Queensland: 3% ethanol (increasing to 4% in 2018), planned introduction of biodiesel mandates (0.5%) in 2017	Current ethanol mandate set to 27% in March 2015 10% biodiesel mandate 2018
Legislation for biofuels/renewable fuels		RenovaBio program in 2018 – New government program to support development of lower carbon biofuels. RenovaBio sets annual national carbon reduction targets. Biofuel producers receive a number of credits (CBio) depending on fuel production volumes and production efficiencies (carbon intensity), based on fuel distributor's individual targets.
Incentives for advanced biofuels	Government provides funding by Australian Renewable Energy Agency for developing advanced biofuel	
Taxes for fuels	Fossil and imported fuels 0.395 AUD l^{-1} Ethanol 0.026 AUD l^{-1}; increasing to ⅓ of the excise rate for petrol by 2030 Biodiesel 0.013 AUD l^{-1} increasing to ½ of the excise rate for diesel by 2030	20% import tax for ethanol is cut down till 2021 For biodiesel imports occurs a 14% tax rate The tax rate for national sold fuel vary from state to state and also depends on feedstock and producer type Petrol Tax (state VAT): 25% to 31% Ethanol Tax (state VAT). 12% to 27% Diesel (federal tax): 248 to 298 BRL m^{-3} Biodiesel (federal tax): 0 to 148 BRLm^{-3}
Energy infrastructure		
Vehicles – Legislation for vehicle emission standards and fuel efficiency	Voluntary target of 6.8 l per 100 km for fuel consumption for new passenger cars (agreed by government and industry) Euro 5 standard is in force for HDV since 2010 and for petrol passenger cars since 2013. The introduction of Euro 6 is still discussed. (AuGov 2016)	The current vehicle emission standard PROCONVE L6 particularly meets Euro 4. The improvement of fuel efficiency is voluntary but heavily supported by large tax and fee discounts for manufacturer meeting the aims of the INOVAR-Auto program. The program will end in 2017 and leads to a 12% fleet improvement in fuel efficiency for light vehicles compared to 2012. (ICCT 2015a)
Incentives for purchasing ZEV/LEV		

References: USDA GAIN Report "Australia Biofuel Annual 2016" USDA GAIN Report "Brazil Biofuel Annual 2016"

Table 12: Task 39 members – non-EU countries

Topic	New Zealand	South Africa
Energy carrier – RE mandates	New Zealand does not have an obligatory biofuel quota. (Kelly 2016)	Compulsory fuel blending mandates are in force since 2015 Bioethanol 2%, Biodiesel 5%
Legislation for biofuels/renewable fuels	The maximum blend rate for ethanol and biodiesel in standard fuels is limited to 10% and 5%. (Kelly 2016)	
Incentives for advanced biofuels		
Taxes for fuels	Petrol: 0.67284 NZD l^{-1} Diesel: 0.53 NZD l^{-1} Petrol/Ethanol blend: 0.0776 NZD l^{-1} + 0.59524 NZD l^{-1} Petrol (NZGOV 2016)	Full fuel tax exemption for bioethanol and a 50% tax reduction for biodiesel
Energy infrastructure	Less than 150 charging stations for EV were installed by the end of 2015	
Vehicles – Legislation for vehicle emission standards and fuel efficiency	All types of new LDV and HDV have to fulfill Euro5/IV emission standard since early 2016. (NZGOV 2012)	
Incentives for purchasing ZEV/LEV	Road user charge exemption for Electric Vehicles until 2020 (Kelly 2016)	

Reference for South Africa: Mandatory blending of biofuels with petrol and diesel (Government Gazette 2013/09/30)

Topic	South Korea
Energy carrier – RE mandates	
Legislation for biofuels/renewable fuels	
Incentives for advanced biofuels	
Taxes for fuels	
Energy infrastructure	
Vehicles – Legislation for vehicle emission standards and fuel efficiency	Emission standards are regulated by Euro 6 for diesel and California´s Non Methane Organic Gases for petrol. The fuel efficiency LDVs was set to 16.7 km l^{-1} and is proposed to be 24.1 km l^{-1} (light trucks 14.1 km l^{-1}) in 2020. (ICCT 2015b)
Incentives for purchasing ZEV/LEV	

Table 14: IEA AMF Annex 52 – countries not covered by preceding tables

Topic	Israel	Thailand
Energy carrier – RE mandates	The share of alternative fuels (e.g. EV, LPG, biofuel) in HDV fleets has to be raised to 3% in 2020 (IMEP 2015)	The mandate for Biodiesel was currently reduced to 5% due to a lack of local available feedstock. The long time aim until 2036 is a B20 blend. No mandatory ethanol blend is in force. The latest national Energy Plan set a goal of 25% biofuel share in 2036.
Legislation for biofuels/renewable fuels		E20/E85 are heavily promoted by price incentives (20% to 40% price advantage).
Incentives for advanced biofuels		
Taxes for fuels	Approximately 65% of the fuel pump price is made by VAT and excise tax (GLOBES 2016)	
Energy infrastructure		1000 charging stations to be built until 2036
		Thai Government boosts availability of E85 with a subsidy for station owners.
Vehicles – Legislation for vehicle emission standards and fuel efficiency	Israel is mainly following the EU (e.g. Euro 5) and US-standards for new cars. Existing fleets of HDV have to meet at least Euro 4 limits by 2018. Fuel meets the Euro 5 Norm (< 10 ppm S) (IMEP 2016)	Euro 4 emission standards in force (UNEP 2016)
Incentives for purchasing ZEV/LEV	Israel´s purchase tax for cars is based on a "Green Grade" (pollution) Index. The maximum tax is set to 83% and it can be lowered down to 10% for ZEV. (OECD 2016)	The Thai government grants a reduced excise tax (17% or 14% instead of 30%) for cars with a E20/85 ready engine with less than 1.3 l displacement and 5 l per 100 km.

References: USDA GAIN Report "Thailand Biofuel Annual 2016"

3   Fuel regulations and fuel standards

3.1   Environmental aspects of fuel regulation

With respect to fuel specifications, relating to properties and market approval of fuels, it should be kept in mind that regulation not only targets technical issues such as compatibility with machinery, gaskets/tubing, vehicle drivability and exhaust gas quality. Other important issues are environmental safety and public health, which can be affected by transport, storage and combusting of fuels. Possible hazards associated with fuels are addressed in existing directives and laws, on which some notes are given here.

Since 1998 it has been determined in the European Union (EU) fuel quality directive FQD 98/70/EC (European Parliament and the Council 1998) that some proven or suspected hazardous fuel components are to be monitored in the environment and that, in case of negative effects, distributive restrictions or maximum contents or even bans may be imposed. This for example applies to (ethyl-)benzene, toluene and the three xylene isomers (also known as BTX class) constituents and, in a 2009 amendment, also to metalloid additives such as methyl-cyclopentadienyl-manganese-tricarbonyl (MMT), with general concern expressed regarding any metalloid additives. For MMT, a limitation has been enforced since January 1, 2014 allowing a maximum content of 2 mg/L. This limit is reflected in both gasoline and diesel specifications (EN 228, EN 590, prEN 16734) and in standard procedures for laboratory analysis (EN 16135, EN 16136, EN 16576).

The EU air quality directive 2008/50/EC (European Parliament and the Council 2008) and respective emissions control acts demand precautions against emissions of ozone precursors from engine operation and fuel handling. To this end, a set of about 30 volatile aliphatic, olefinic and aromatic hydrocarbons plus formaldehyde plus NO_X is referred to in annex X for ozone precursor control. The volatile hydrocarbons (HC) mentioned are liberated both by evaporative losses from uncombusted fuels and by engine exhaust, the latter adding formaldehyde (among other carbonylics) and NO_X. Measurement of volatile HC ozone precursors in ambient air for emissions control requires GC/MS methods to achieve compound-specific results. For research purposes, GC/MS is likewise applied in some exhaust gas investigations, though routinely gaseous concentrations are reported as total hydrocarbons (THC) including methane and non-methane [volatile] hydrocarbons (NM[V]HC). Determination of methane emissions data is also required for liquid gas (LG) vehicles in the EU.

European legislation on soil and water protection generally requires controlling and monitoring hazardous substances and considers implementing restrictions if evidence for adverse effects exists. This of course applies to fuels or fuel components since related infrastructure and motor traffic span large areas, with corresponding ubiquitous contamination risks resulting from leakage, misuse, accidental or intentional spillage. In fact, in urban settings or along congested traffic zones fossil hydrocarbons and light aromatic compounds give rise to typical contamination patterns in soil, water and groundwater. Aliphatic ethers like Methyl-tert-butylether (MTBE), used as an oxygenate and anti-knocking agent in some gasoline fuels, have a significantly higher water solubility than hydrocarbons, which is the reason for their fast infiltration into vadose zones and aquifers. Due to their low degradability in soil and water environments under suboxic or anoxic conditions, these ethers can remain in aquatic environments unaltered for long periods and contaminate groundwater. MTBE pollution of groundwater meanwhile has become a matter of concern after widespread contamination of drinking water resources in the US following MTBE introduction as fuel component caused by leaking underground storage tanks (Weaver et al., 2010). Though groundwater contaminations have occurred in Europe too, fuel borne MTBE is not regarded as a severe problem, because strict regulations on handling and storage facilities are considered sufficient for prevention of unwanted releases.

Niven (2005) gives a detailed discussion on environmental aspects of fuel safety in general and examines fuel ethanol in particular. Swick et al. (2014) published a compendium on legal and practical issues of gasoline risk management. Yang, Hollebone et al. (2015) provide an example of photooxidative degradation studies of diesel and biodiesel blends in freshwater, which is just one approach to remedial action to be pursued in the case of accidental fuel releases into the environment.

3.2 Access to standards: fuel specifications and laboratory test methods

There are two categories of regulations pertaining to fuel properties:

- Specification of fuels – specific parameters (threshold values) to be met by fuels, as determined in annexes of governing laws or by virtue of corresponding technical standards (e.g. EN 14214/ASTM D 6751 for Biodiesel);

- Specification of analysis procedures – detailed instructions for fuel sampling and analysis to be met by executing laboratories, including quality control and documentation.

The former item, specification of fuels, will be substantiated in Section 3.3 including Table 15, while detailed reference on the latter, specification of analysis procedures, is outside of this study's scope. This said, comments on procedures for parameter determination are provided where appropriate.

For information and download of standards (fuel specifications and test methods), the following online resources are available:

Beuth Verlag, english version:

> http://www.beuth.de/en

Energy Institute:

> http://publishing.energyinst.org/ip-test-methods

Enter method name or search item, or refer to complete listing at:
> https://www.energyinst.org/filegrab/?ref=4985&f=list-of-ip-test-methods.pdf
> (List of IP test methods, panels responsible for them and corresponding BS 2000, EN, ISO and ASTM methods)

ASTM and mirror sites:

> https://www.astm.org/
> http://www.techstreet.com/
> http://standards.globalspec.com/

To identify specific ASTM standards related to fuels and fuel testing (Section 5: Petroleum Products, Lubricants, and Fossil Fuels), refer to document listings provided by the ASTM Bookstore or Japanese Standards Association (JSA) webstore:

> https://www.astm.org/BOOKSTORE/BOS/TOCS_2016/05.01.html
> https://webdesk.jsa.or.jp/link/webstore/Book/html/jp/ad/contents/astm16_0501.pdf
> (Petroleum Products, Liquid Fuels, and Lubricants (I): C1234-D3710)

> https://www.astm.org/BOOKSTORE/BOS/TOCS_2016/05.02.html

http://webdesk.jsa.or.jp/link/webstore/Book/html/jp/ad/contents/astm16_0502.pdf
(Petroleum Products, Liquid Fuels, and Lubricants (II): D3711-D6122)

https://www.astm.org/BOOKSTORE/BOS/TOCS_2016/05.03.html
http://www.webstore.jsa.or.jp/webstore/Book/html/jp/ad/contents/astm16_0503.pdf
(Petroleum Products, Liquid Fuels, and Lubricants (III): D6138-D6971)

https://www.astm.org/BOOKSTORE/BOS/TOCS_2016/05.04.html
https://webdesk.jsa.or.jp/link/webstore/Book/html/jp/ad/contents/astm16_0504.pdf
(Petroleum Products, Liquid Fuels, and Lubricants (IV): D6973-D7755)

https://www.astm.org/BOOKSTORE/BOS/TOCS_2016/05.05.html
https://webdesk.jsa.or.jp/link/webstore/Book/html/jp/ad/contents/astm16_0505.pdf
(Petroleum Products, Liquid Fuels, and Lubricants (V): D7756-latest; Combustion Characteristics; Manufactured Carbon and Graphite Products)

https://www.astm.org/BOOKSTORE/BOS/TOCS_2016/05.06.html
https://webdesk.jsa.or.jp/link/webstore/Book/html/jp/ad/contents/astm16_0506.pdf
(Gaseous Fuels; Coal and Coke; Bioenergy and Industrial Chemicals from Biomass; Catalysts)

URLs are valid for 2016 documents and will be updated in spring each year.

Outlines and relevancy of ASTM test methods in the field of petroleum/lubricant analysis, cross referenced to equivalent standards from other standards organizations, are compiled in the "Guide to ASTM Test Methods for the Analysis of Petroleum Products and Lubricants" by Nadkarni (2007), see

https://www.astm.org/DIGITAL_LIBRARY/MNL/SOURCE_PAGES/MNL44-2ND.htm,

additionally available online at

http://197.14.51.10:81/pmb/CHIMIE/petrochimie/Guide%20to%20Astm%20Test%20Methods%20for%20the%20Analysis%20of%20Petroleum%20Products%20and%20Lubricants,%20Second%20Edition.pdf,

or a comprehensive report on international fuel standards, their relevance and current issues of necessary adaptations has been compiled for the Australian Government, Department of the Environment, by Hart Energy Research & Consulting (2014).

3.3 Overview on fuel standards (specifications)

Main commercially available fuels for road vehicles

- fossil diesel with up to 7% of fatty acid methyl ester (FAME, biodiesel)
- diesel blends with max. 10% [EU] or 20% [US] v v^{-1} biodiesel
- neat biodiesel
- common fossil gasoline with up to 5%, 10% or 20% of ethanol
- gasoline blends with 85% v v^{-1} methanol or ethanol (flex-fuel vehicles)
- hydrous ethanol

For these fuels, the following regulations (standards) apply:

Fossil diesel with minor contents of biodiesel

(Column [1] in Table 15)

- **DIN EN 590:2017-10**: Automotive fuels - Diesel - Requirements and test methods.

Different fuel classes according to climatic conditions (cold properties, viscosity, density, cetane, volatility).

> https://www.beuth.de/en/standard/din-en-590/278413784

- **ASTM D975-18**: Standard Specification for Diesel Fuel Oils.

Different fuel grades according to engine requirements (sulfur, volatility; speed/load alteration).

> https://www.astm.org/Standards/D975.htm

Diesel blends with max. 10% [EU] or 20% [US] v v^{-1} biodiesel (FAME)
(Column [2] in Table 15)

- **DIN EN 16734:2016-11**: Automotive fuels – Automotive B10 diesel fuel – Requirements and test methods.

Different fuel classes according to climatic conditions (cold properties, viscosity, density, cetane, volatility).

> http://www.beuth.de/en/standard/din-en-16734/250352855

- **ASTM D7467-18**: Standard Specification for Diesel Fuel Oil, Biodiesel Blend (B6 to B20).

Different fuel grades according to sulfur content with reference to 6 out of 7 base fuel grades in D975 (Grade No. 4-D—heavy distillate fuel not allowed).

> https://www.astm.org/Standards/D7467.htm

Recently, in the EU the new standard DIN EN 16709 for diesel-biodiesel blends with volumetric biodiesel contents up to 20 or 30% was introduced. It is under examination and only valid for captive fleet vehicles.

Neat biodiesel
(Column [3] in Table 15)

- **DIN EN 14214:2014-06**: Liquid petroleum products – Fatty acid methyl esters (FAME) for use in diesel engines and heating applications – Requirements and test methods.

Different fuel classes according to climatic conditions (CFPP). Complement for biodiesel intended as blend component: different classes according to climatic conditions (CP, CFPP, monoglyceride content).

> http://www.beuth.de/en/standard/din-en-14214/197713876

- **ASTM D6751-15ce1**: Standard Specification for Biodiesel Fuel Blend Stock (B100) for Middle Distillate Fuels.

Different fuel grades according to engine requirements (sulfur, low-temp., sensitivity to glycerides).

> https://www.astm.org/Standards/D6751.htm

Compilations of specification parameters and test methods for biodiesel and biodiesel blends:

DieselNet.com information page "Biodiesel Standards & Properties",

https://dieselnet.com/tech/fuel_biodiesel_std.php ;

U.S. Department of Energy – Alternative Fuels Data Center, document "ASTM Biodiesel Specifications",

http://www.afdc.energy.gov/fuels/biodiesel_specifications.html ;

Short summary of regulated European biodiesel parameters (in English), their relevance and applicable test methods is provided by German Association Quality Management Biodiesel (AGQM),

http://www.agqm-biodiesel.de/files/5813/6066/8476/AGQM_0183_Merkblatt-Analytik-eng_080113.pdf

Fossil gasoline with minor contents of ethanol
(Column [5] in Table 15)

- **DIN EN 228:2017-08**: Automotive fuels - Unleaded petrol - Requirements and test methods.

Different fuel classes according to total oxygen content (3.7% and 2.7%) and volatility.

https://www.beuth.de/en/standard/din-en-228/273470473

- **ASTM D4814-18a**: Standard Specification for Automotive Spark-Ignition Engine Fuel.

Different fuel grades according to engine requirements, climatic conditions and elevation (volatility, water tolerance, driveability index).

https://www.astm.org/Standards/D4814.htm

High-level blends of ethanol in fossil gasoline
(Column [6] in Table 15)

- **DIN 51625:2008-08**: Automotive fuels – Ethanol Fuel – Requirements and test methods.

http://www.beuth.de/en/standard/din-51625/109584985

- **DIN CEN/TS 15293:2011-04**; **DIN SPEC 91220:2011-04**: Automotive fuels – Ethanol (E85) automotive fuel – Requirements and test methods (parameter specs. from this Tech. Ref. were used in Table 15).

http://www.beuth.de/en/technical-rule/din-cen-ts-15293/128346437

- **DIN EN 15293:2017-06 – Draft**: Automotive fuels – Ethanol (E85) automotive fuel – Requirements and test methods. This draft version closely follows respective 2011 Technical Reference, except: max values for ethers, copper and gum omitted, max. allowable sulfate reduced to 2.6 mg/kg.

https://www.beuth.de/en/draft-standard/din-en-15293/271517131

- **ASTM D5798-17**: Standard Specification for Ethanol Fuel Blends for Flexible-Fuel Automotive Spark-Ignition Engines.

https://www.astm.org/Standards/D5798.htm

Fuels for regional or research application and blending purposes

Among other important liquid fuels that have minor market share and/or are intended for blending purposes instead of direct customer (end-user) sale, the following have been regulated by specifications:

- fuels from hydro-thermochemical processing (HVO, BTL, GTL)
- high-level blends of methanol in fossil gasoline
- neat ethanol for gasoline blending
- neat butanol for gasoline blending
- DME

Fuels from hydro-thermochemical processing (HVO/HEFA, BTL, GTL)
(Column [4] in Table 15)

- **DIN EN 15940:2016-09:** Automotive fuels – Paraffinic diesel fuel from synthesis or hydrotreatment – Requirements and test methods.

     http://www.beuth.de/en/standard/din-en-15940/245465614

An attempt had been made to establish a new ASTM-standard for paraffinic diesel fuel, which has not yet been successful (see note of a discussion meeting report of ASTM Committee D02, 2013,

     https://www.astm.org/COMMIT/D02%20New%20Specifications%20Forum%20Summary.pdf,

on p. 4 it says: "Paraffinic diesel fuel / Paraffinic middle distillate – 6 ballots failed; proponents discontinued their efforts for an ASTM specification").

In contrast, synthetic paraffinic fuels of FT-, HEFA-, and SIP-type (SIP: Synthesized Iso-Paraffinic) have entered the aviation fuel market as blend components with according standardization in ASTM D 1655 (Specification for Aviation Turbine Fuels) and ASTM D 7566 (Standard Specification for Aviation Turbine Fuel Containing Synthesized Hydrocarbons).

High-level blends of methanol in fossil gasoline

- No applicable fuel standard among EU legislation yet
- **ASTM D5797-17:** Standard Specification for Methanol Fuel Blends (M51-M85) for Methanol-Capable Automotive Spark-Ignition Engines.

     http://www.astm.org/Standards/D5797.htm

In the EU, methanol fuels to date have not gained market entrance, partly due to its comparatively low energy content and poor acceptance because of its acute toxicity. Markets with considerable share of higher methanol blends like the US and China have elaborated industrial standards and/or recommendations on pure methanol furnished for vehicle fuel purposes.

Hydrous ethanol
(Column [7] in Table 15)

Brazil standards for hydrous ethanol fuel can be found in

An English translation is available in

http://www.itecref.com/pdf/Brazilian_ANP_Fuel_Ethanol.pdf

Two standards are specified:
Hydrated ethanol fuel (EHC) and hydrated ethanol fuel premium (EHCP) destined for direct use in internal combustion engines.

Neat ethanol for gasoline blending

- **DIN EN 15376:2014-12**: Automotive fuels – Ethanol as a blending component for petrol – Requirements and test methods.

  http://www.beuth.de/en/standard/din-en-15376/205428341

- **ASTM D4806-17**: Standard Specification for Denatured Fuel Ethanol for Blending with Gasolines for Use as Automotive Spark-Ignition Engine Fuel.

  https://www.astm.org/Standards/D4806.htm

Neat ethanol is not intended for direct sale to fuel end-users.

Neat butanol for gasoline blending

- No applicable fuel standard among EU legislation yet

- **ASTM D7862-17**: Standard Specification for Butanol for Blending with Gasoline for Use as Automotive Spark-Ignition Engine Fuel.

  https://www.astm.org/Standards/D7862.htm

The following Table 15 compiles fuel specification standard parameters regulated by EN or ASTM for liquid fuel types mentioned above that are commonly available. This table is intended to provide a correlative overview of applicable parameters that must be included in fuel specifications. Details for climate-dependent parameters like cold-flow properties and volatility classes have been omitted for the sake of clarity; see footnotes for more parameters specific to single fuel types.

Table 15: Standards for common commercial liquid fuels (customer sale) and their main parameters

EN/DIN: status 2016;
ASTM: (1)/(5)/(6) status 2014, HartEnergy Report 2014; (2)/(3) status 2013/2015, Alternative Fuels Data Center, U.S. Dept. of Energy
min = minimum allowed values; max = maximum allowed values; n.spec. = not specified

Column	[1]	[2]	[3]	[4]	[5]	[6]	[7]
Fuel	Fossil diesel up to 7% of FAME	Diesel blends with max.10 or 20% v v^{-1} FAME	Neat FAME	Diesel from hydro-/thermo-chemical processing (HVO/HEFA, XTL)	Fossil gasoline with minor contents of ethanol	High-level blends of ethanol in fossil gasoline	Hydrated ethanol fuel Premium (Brazil)
Parameter [unit]	EN 590 ASTM D975	EN 16734 ASTM D7467	EN 14214 ASTM D6751	EN 15940 [class 1 // 2] no ASTM	EN 228 ASTM D4814	DIN 51625 EN 15293[2011] ASTM D5798	
Cetane number [-]	51 min 40 min	51 min 40 min	51 min 47 min	70 // 51 min n.spec.			

27

Column	[1]	[2]	[3]	[4]	[5]	[6]	[7]
Fuel	Fossil diesel up to 7% of FAME	Diesel blends with max.10 or 20% v v⁻¹ FAME	Neat FAME	Diesel from hydro-/thermo-chemical processing (HVO/HEFA, XTL)	Fossil gasoline with minor contents of ethanol	High-level blends of ethanol in fossil gasoline	Hydrated ethanol fuel Premium (Brazil)
Cetane index [-]	46 min / n.spec.	46 min / 40 min					
Octane number Research [-]					95 min (AKI)		
Octane number Engine [-]					85 min (AKI)		
Density (15 °C) [kg m⁻³]	820-845 / n.spec.	820-845 / n.spec.	860-900 / n.spec.	765-800 // 780-810 / n.spec.	720-775 / n.spec.	760-800 / n.spec.	805.2-811.2 / 799.7-802.8
PAH [% m m⁻¹]	8 max / n.spec.	8 max / n.spec.					
Total aromatics	n.spec. / 35% v v⁻¹ max	n.spec. / 35% v v⁻¹ max		1.1% m m⁻¹ max / n.spec.	35% v v⁻¹ max / n.spec.		
Sulfur [mg kg⁻¹] (ppm)	10 max / 15 max	10 max / 15 max	10 max / 15 max	5 max / n.spec.	10 max / 80 max	10 max. / 80 max	Report
Phosphorus				4 mg kg⁻¹ max / n.spec.; 10 mg kg⁻¹ max / n.spec.	n.spec. / 1.3 mg l⁻¹ max	0.15 mg l⁻¹ max / 1.3 mg l⁻¹ max	
Manganese [mg l⁻¹]	2 max / n.spec.	2 max / n.spec.		2 max / n.spec.	2 max / n.spec.	n.spec. / n.spec.	
Lead [mg l⁻¹]					5 max / 13 max	n.spec. / 13 max	
Flash point [°C]	55 min / 52 min	55 min / 52 min	101 min / 93 (130) min	55 min / n.spec.			
Carbon residue of 10% dist. residue [% m m⁻¹]	0.30 max / 0.15/0.35 max	0.30 max / 0.35 max	n.spec. / 0.05* max	0.30 max / n.spec.			
(sulfated) Ash [% m m⁻¹]	0.01 max / 0.01 max	0.01 max / 0.01 max	0.02 max / 0.02 max	0.01 max / n.spec.			
Water [mg kg⁻¹]	200 max / n.spec.	200 max / n.spec.	500 max / n.spec.	200 max / n.spec.		4,000 max / 10,000 max	7.5% / 4.5%
Total contam. or Water + Sediment [mg kg⁻¹] or [% v v⁻¹]	24 max / 0.05 max	24 max / 0.05 max	24 max / 0.05 max	24 max / n.spec.			
Evapor. residue (gum, washed) [mg (100 ml)⁻¹]					5 max / 5 max	5 max / 5 max	5 max
Copper corros. (3 h/50 °C) [class]	1 / 3	1 / 3	1 / 3	1 / n.spec.	1 / 1	1 / n.spec.	
FAME content [% m m⁻¹]	7 max / 5 max	10 max / 6-20	96.5 min / n.spec.				
Ethanol [% v v⁻¹]					5 // 10 / n.spec.	70-85 / 51-83	92.5-94.6 / 95.5-96.5

Column	[1]	[2]	[3]	[4]	[5]	[6]	[7]
Fuel	Fossil diesel up to 7% of FAME	Diesel blends with max.10 or 20% v v^{-1} FAME	Neat FAME	Diesel from hydro-/thermo-chemical processing (HVO/HEFA, XTL)	Fossil gasoline with minor contents of ethanol	High-level blends of ethanol in fossil gasoline	Hydrated ethanol fuel Premium (Brazil)
Oxidation stability ** [g m^{-3}] or [h]	25 g m^{-3} max or 20 h min n.spec.	25 g m^{-3} max or 20 h min 6 h min	8 h min 3 h min	25 g m^{-3} max or 20 h min n.spec.	6 h min 4 h min	6 h min n.spec.	
Lubricity at 60 °C [µm]	460 max 520 max	460 max 520 max	n.spec. n.spec.	460 max n.spec.			
Viscosity at 40 °C [mm^2 s^{-1}]	2.0 to 4.5 1.9 to 4.1	2.0 to 4.5 1.9 to 4.1	3.5 to 5.0 1.9 to 6.0	2.0 to 4.5 n.spec.			
Txx or Final B. P. [°C]	T95: 360 max T90: 282 to 338 max	T95: 360 max T90: 343 max	T90: 360 max T90: 360 max	T95: 360 max n.spec.	FBP: 210 max FBP: 225 max	n.spec. n.spec.	
CFPP temprt. or Cold Soak Filt. [°C] or [s]	+5 to -20 °C max n.spec.	+5 to -20 °C max n.spec.	+5 to -20 °C max 360 s max	+5 to -20 °C max n.spec.			
CFPP arctic. or Cold Soak Filter. [°C] or [s]	-20 to -44 °C max n.spec.	-20 to -44 °C max n.spec.	-20 to -44 °C max **** 200 s max	-20 to -44 °C max n.spec.			
CP arctic *** [°C]	-10 to -34 °C max n.spec.	-10 to -34 °C max n.spec.	-10 to -34 °C max **** Report	-10 to -34 °C max n.spec.			
Specific parameters for distinct fuel type	*(+) diesel*		*(+) biodiesel*		*(+) gasoline low EtOH*	*(+) gasoline high EtOH*	

* From 100% of distillation sample; ** depending on method applied
*** Other adjustments with density, viscosity, cetane, distillation
**** Deviating values if biodiesel is to be used as blend component

Specific parameters for distinct fuel type, either EN or ASTM:

(+) Diesel: polyaromatics, distillation E250-350/T90-95
(+) Biodiesel: acid number; iodine value; content of methanol, linolenic methyl ester, polyunsaturated FAME with 4+ C=C-bonds, mono-/di-/triglycerides, free/total glycerol, Na+K, Ca+Mg, P
(+) Gasoline low ethanol: olefinics, aromatics, benzene, total oxygen, methanol, ethanol, other lower alcohols, ethers, vapor pressure/distillation E70-100-150/T10-50-90; Anti-Knock-Index, Vapor Lock Index, driveability Index DI, silver corrosion.
(+) Gasoline high ethanol: acidity (acetic acid), methanol, conductivity, higher alcohols, inorganic chlorine, sulfate [EN: ethers, gum and copper omitted in recent update]; similar to spec. for neat ethanol for blending (EN 15376). ASTM: unwashed gum

Liquefied (pressurized) gaseous fuels

- Natural gas and biogas (LNG, LBG, LBM; mostly methane)

- Liquefiable fossil hydrocarbon gas (LPG, Autogas; mostly propane/butane)

- DME

Dimethyl ether (DME)

Preliminary drafts of standards for DME showed slightly varying parameter values in the past, cf. AMF Annex 47, Reconsideration of DME Fuel Specifications for Vehicles,

http://iea-amf.org/content/projects/map_projects/47;

or Fleisch et al. (2012).

Meanwhile, ISO, ASTM and JSA have established valid standards for DME fuel application, but no corresponding EN standard has been established yet.

ISO: http://www.iso.org/iso/iso_technical_committee.html?commid=47414, search topic "DME" provides standards on fuel quality and determination of DME parameters.

- **ISO 16861:2015** – Petroleum products – Fuels (class F) – Specifications of dimethyl ether (DME).

  https://www.iso.org/standard/57835.html

- **ASTM D7901-14b**: Standard specification for Dimethyl Ether for Fuel Purposes.

From standard outline: "1.1 This specification covers dimethylether (DME) for use as a fuel in engines specifically designed or modified for DME and for blending with liquefied petroleum gas (LPG)".

  https://www.astm.org/Standards/D7901.htm

Table 16 provides standard parameter settings applicable to hydrocarbon gaseous fuels and DME in a comparative overview.

Table 16: Standards for gaseous hydrocarbon fuels, DME and their main parameters

EN/DIN: status 2016;
ASTM: (1)/(2)/(3) status 2014, HartEnergy Report 2014; ISO: (4) status 2013/2014, AMF Fuel Info respectively Annex 47
Min = minimum allowed values, max = maximum allowed values; n.spec. = not specified
see table footnotes for further remarks

Column	[1]	[2]	[3]	[4]
Fuel	Natural gas and biomethane – transport, gas network	Automotive fuels – LPG	DME	DME
Parameter [unit]	prEN 16723-2 ASTM n.spec.	EN 589 ASTM D1835	n.spec. ASTM D 7901-14a	n.spec. ISO/DIS 16861 draft
Octane number Engine [-]		89 min n.spec.		
Methane number [-]	65/80 min n.spec.			
Lower heating value [MJ kg^{-1}]	39 min (automot.) n.spec.			
Lower Wobbe-Index (H) [MJ m^{-3}]	41.9 min/49 max n.spec.			
Lower Wobbe-Index (L) [MJ m^{-3}]	35.0 min/40.8 max n.spec.			
Vapor pressure [kPa]		1,550 max (40 °C) 1,434 max (37.8 °C)	n.spec. 758 max (37.8 °C)	
Hydrocarbon dew point [°C]	-2 max n.spec.	n.spec. n.spec.		
T95 [°C]		n.spec. -38.8 max		
Evaporation resid. [mg kg^{-1}]		60 max 0.05/100 ml max	n.spec. 0.05/100 ml max	n.spec. 0.007/100 ml max
Oil stain observation [-]		n.spec. pass	n.spec. pass	
Lubricity at 60 °C [μm]			n.spec. consult	

Column	[1]	[2]	[3]	[4]
Fuel	Natural gas and biomethane – transport, gas network	Automotive fuels – LPG	DME	DME
Viscosity at 40 °C [mm² s⁻¹]				
Hydrogen [mol-%]	2 max / n.spec.			
Oxygen [mol-%]	1 max / n.spec.			
Total dienes (incl. 1,3 butadiene) [mol-%]		0.5 (5% v v⁻¹ propylene)		
Hydrocarbons ≤ C₃ [% m m⁻¹]				n.spec. 0.05
H₂S + COS [mg m⁻³]	5 max / n.spec.	Not detectable / pass		
Sulfur [mg kg⁻¹; ppm]	10 max (automot.) / n.spec.	50 max (odor.) / 123 max (odor.)	n.spec. 3 max	n.spec. 3 max
Water content/ moisture [-]		pass / pass	n.spec. 0.03% m m⁻¹	n.spec. 0.05% m m⁻¹
Water dew point [°C, at 200 bar]	-10...-30 °C max / n.spec			
Copper corros. (1 h/40 °C) [class]		1 max / 1 max	n.spec. 1 max	
DME content [% m m⁻¹]			n.spec. 98.5 min	n.spec. 99.5 min
Methanol [% m m⁻¹]			n.spec. 0.05 max	n.spec. 0.05 max
CO₂ [% m m⁻¹]				n.spec. 0.01 max
CO [% m m⁻¹]				n.spec. 0.05 max
Methylformate [% m m⁻¹]				n.spec. 0.2 max
Specific parameters for distinct fuel type	*(+) LNG/LBG*	*(+) LPG*		

Specific parameters for distinct fuel type, either EN or ASTM:

(+) LNG/LBG: Total silicon; compressor oil; dust/particulate contamination; informative annex on odoration. Older German standard DIN 51624 for natural gas further specified density; methane 80% min.; C_2 to C_{6+} hydrocarbon, nitrogen and CO_2 maximum contents

(+) LPG: Vapor pressure min. 150 kPa at temperature -10 °C … +20 °C class A … E; smell unpleasant and specific (odoration)

4 Technical background of selected fuel properties

Beside sustainability, technical feasibility and security of supply, the chemical nature of fuel components and their interdependence with engine/vehicle functionality have to be addressed to attain a sound basis for selection of advanced fuels. Since a broad range of oxygen-free as well as oxygenated organic compound classes have proved suitable for use in modern engines, premature exclusion of particular substances should be avoided. This is especially important with respect to ongoing research on optimizing engine design and combustion kinetics, which has the potential to favorably exploit different fuel types, including distinct biodiesel blends. Moreover, fossil fuels are regarded as an indispensable part of the fuel supply on a short- to mid-term perspective. Consequently, compatibility and possible interactions between traditional fuels and advanced fuels remains a matter of concern. At the same time, promising dual-fuel concepts are being developed, whereby desired functionalities from diverse fuel types can be combined to achieve targeted performance levels.

Within the area of fuel supply and logistics, two aspects of upcoming changes in fuel type should be highlighted. First, the degree of similarity between common fossil fuels and any alternative fuels has consequences for future usability of existing infrastructure and technical equipment. Second, the amount of resources and energy required to produce a highly upgraded, energy-rich fuel strongly influences its production cost. From an economic viewpoint it is advisable to adapt fuel chemistry to existing techniques (engines), which is the overall idea underlying prioritization of drop-in fuels (IEA 2014). According to this concept, carbonaceous feedstock from biogenic sources, wastes or secondary petrochemical sources is converted via successions of thermochemical and/or biochemical procedures into an upgraded product, i.e., highly saturated hydrocarbons similar to and compatible with common petroleum-based fuels. Hydrogen demand for this pathway is inherently high (or conversely carbon yield is low if oxygen is lost as CO_2), since oxygen content of starting materials will ultimately need to be reduced to zero. By re-building fossil-derived templates, such fuel types to some extent become stipulated for future engine use. This can only be future-proof if installed production plant capacities together with sufficient amounts of hydrogen can be guaranteed and overall process efficiency allows sustainable production. However, seeking the highest degrees of upgrading and fuel energy density actually worsens the overall life-cycle sustainability footprint. Considering the picture of an "energy level staircase" (as depicted on pages 34, 38 and 163 in IEA 2014), it must be checked at which step the disadvantage of "lower" fuel quality (as described by energy content/heating value and combustion characteristics associated with molecular structure) is outweighed by saving resources for further upgrading and, instead, using the fuel for efficient combustion in well suited engines. This has been discussed by Hao et al. (2016) based on a comparative evaluation of well-to-tank and tank-to-wheel analyses applied to traditional gasoline spark ignition (SI) engines running on octane number (ON) fuels with ON<70.

To get a deeper insight into such trade-off implications, the reader is referred to instructive reviews by Westbrook (2013) and Bergthorson & Thomson (2014) on interrelationships between fuel type and engine combustion concepts. The potential of comparatively low-grade fuels is also described by Zhang et al. (2016b) in a recent diesel combustion study on naphtha and ultra low sulfur diesel (ULSD), demonstrating that current knowledge on engine control allows low-emission combustion of lower quality fuels that otherwise would be insufficient in terms of meeting standard market rules. Another interesting example in this respect, at the same time serving as conceptual study for dual-fuel strategies, is presented by Morganti et al. (2015). They explored potential benefits of torque-dependent, additional methanol dosage from a separate fuel tank in two common gasoline engines equipped with respective auxiliaries. The vehicles were run on low-octane gasoline (termed "naphtha A/B") in low- and medium-torque driving situations, while methanol as high-octane make-up component was injected under severe driving conditions ("octane-on-demand"). Evaluation of results from US06 cycle test runs showed reductions up to 27% in CO_2 emissions. This octane-on-demand concept has gained recognition recently and can typically be accomplished by, though it is not restricted to, dual-fuel or modifications to on-board fuel infrastructure; for further information, see the investigation by Rankovic et al. (2015) and the Ph.D. dissertation by Jo (2016). Continued progress in the development of improved engine combustion technologies will further broaden the range of possible applications.

For a basic overview on fuel properties the reader is referred to the AMF fuel information site, available online at

> http://www.iea-amf.org/content/fuel_information/fuel_info_home.

The present chapter's focus is on advanced biofuels such as HVO, BTL, DME, OME, alcohols, and biodiesel (FAME). For selected fuel properties updating and complementing these data with additional information and results from recent literature are summarized.

Institutions and information portals acting at global, regional or national levels accomplish controlling and reporting on available fuel qualities and their respective specifications. Information thus provided includes tabulated data, results from laboratory analyses or round robin campaigns as well as supplementary material gained from investigations on special issues. See for example resources provided online by:

WWFC – Worldwide Fuel Charter, 5th ed. 2013:

> http://www.acea.be/uploads/publications/Worldwide_Fuel_Charter_5ed_2013.pdf

Alternative Fuels Data Center (US Department of Energy, http://www.afdc.energy.gov/):

> http://www.afdc.energy.gov/fuels/emerging.html
> http://www.afdc.energy.gov/fuels/fuel_properties.php
> http://www.afdc.energy.gov/fuels/properties_notes.html

NREL Technical Report NREL/TP-5400-52905 by T.M. Alleman, "National 2010-2011 Survey of E85":

> https://crcao.org/reports/recentstudies2012/E-85-2/CRC%20Project%20E-85-2%20Final%20Report.pdf

AGQM (German Association Quality Management Biodiesel e.V.):

> http://www.agqm-biodiesel.de/en/ (English version), report for 2015:
> http://www.agqm-biodiesel.de/files/4214/6978/3315/AGQM_Quality_Report_2015.pdf

Fuel manufacturers and blenders hold some relevant publications too, see e.g.

Chevron 2007 – Diesel Fuels Technical Review:

> https://www.chevron.com/-/media/chevron/operations/documents/diesel-fuel-tech-review.pdf

Neste 2016 – Renewable Diesel Handbook:

> https://www.neste.com/sites/default/files/attachments/neste_renewable_diesel_handbook.pdf

Infineum 2014 – Winter Diesel Survey:

https://www.infineum.com/media/80722/wdfs-2014-full-screen.pdf

Greenergy UK – Sales Specifications and Analysis Certificates:

http://www.greenergy.com/uk/technical
http://www.greenergy.com/uk/quality

Methanol Institute (www.methanol.org ; www.methanolfuels.org) – Techn. Bulletin Flammable Liquids:

http://www.methanol.org/wp-content/uploads/2016/07/UsingPhysicalandChemicalPropertiesto
ManageFlammableLiquidHazardsPart1b.pdf

In case of singular chemicals like methanol, deviations in fuel properties would be a matter of side products or impurities. The more complex a fuel is in terms of chemical species and starting material to be processed, the higher possible scatter in properties and reactivities of the fuel product(s). Additional refining steps, quality control and laboratory analysis may be necessary to arrive at desired final fuel specifications.

Most readers will be familiar with the basic importance and relevance of fuel properties, and why these are subject to stringent control by fuel specifications. Therefore, remarks given in this section are not intended to replace a fuel textbook or to provide a comprehensive introduction on this topic. Instead, we point to aspects of fuel descriptors that are relevant in subsequent discussions of chemical reactions and stability as well as health effects and engine emissions. Inspection of fuel specifications might reveal that some presumably important parameters are missing (as, for example, viscosity among gasoline fuels) – which mostly follows from a fuel property falling well within a suitable range. Nevertheless, investigations dedicated to clarify general mechanisms and interdependences will seek analysis data for any fuel property that might influence operability and exhaust gas quality.

Two basic physical bulk properties, density and viscosity, are closely related to proper action of fuel system components like pumps, valves and filters. Exceeding specified parameter limits would negatively affect fuel injection timing and amount. Large differences in density and/or viscosity among fuels would be unfavorable in everyday practice after refilling events, since inhomogeneities from incomplete mixing introduce uncontrolled variability in the quality of fuel actually being delivered to injectors. Local inhomogeneity also increases risks for phase separations in the case of falling temperatures or increasing water content.

Fuel lubricity relates directly to durability of mechanical parts of fuel pumps. Additionally, enhanced wear and abrasion due to insufficient lubricity will subsequently contribute to fuel deterioration, which especially affects fuel circulating in common rail systems. While underlying mechanisms are still subject to engineering research, it can be assumed that both local thermal stress at poorly lubricated surfaces and catalytic action of liberated trace metals contribute to deterioration reactions, adding to adverse oxidative reactions. In order to discern consequences of the latter from those of frictional wear and fretting, test setups for accelerated wear like extreme pressure (EP) testing of lubricating fluids are methods of choice. Hu et al. (2012) for example report on relevant investigations for the oxygenate fuel dimethylfuran, applying ASTM method D2783 for stress exertion and performing product identification by Fourier Transformed Infrared Spectroscopy (FT-IR) and Gas Chromatography/Mass Spectroscopy (GC/MS).

For further information, a review of fuel lubricity fundamentals that also includes a literature survey and assignment of lubricity properties to some fuel constituents is given by Hsieh and Bruno (2015) and a university publication by Möller and deVaal (2014) describes mechanistic concepts and tribology and wear measurements (Möller, V. P.; de Vaal, 2014). Additionally, in a Ph.D. thesis, Langenhoven (2014) demonstrates the influence of trace amounts of water on lubricity testing of model substances related to fuels by close inspection of wear surfaces and application of Raman spectroscopy to detect iron oxide and carbon. A detailed description of methods and mechanistic concepts is provided.

Volatility and vapor pressure of liquid fuel components are of central importance in several respects. In a general sense, liquid fuels belong to semi-volatile organics, which means that they can be kept and transported in liquid state without pressurized containers while upon moderate heating they are volatized. The boiling ranges of common fossil fuels, HVO/HEFA, and synthetic XTL fuels typically are on the order of 30 to 215 °C (gasoline) and 160 to 370 °C (diesel). Constituents occupying the lower end of the boiling range are critical regarding volatilization losses (gasoline), cold-start behavior and ignitability (gasoline, diesel), while the highest boiling components influence the efficacy of fuel atomization to achieve proper combustion of cylinder charge. Reid vapor pressure (RVP) for spark-ignition fuels is determined by low-boiling fuel components and has to be adjusted seasonally (front-end volatility), as determined in respective fuel specifications for climate-dependent parameters.

Production processes for XTL and metathesis fuels offer possibilities to shift product composition towards desired boiling ranges. Other advanced fuels under consideration like lower molecular weight alcohols, ethers and OME-class fuels provide somewhat constrained boiling ranges, or even distinct boiling points as in case of pure compounds. Adjustable boiling characteristics of fuels of course are also an important co-factor in the context of fuel injection strategies for modern engines.

To determine boiling ranges of fuels, it is not necessary to perform common distillation procedures of fuel samples. Applicable laboratory standards make use of simulated distillation (SimDis), which in fact comprises capillary gas chromatography, the eluent retention times of which have been correlated and standardized with reference substance mixtures of known boiling points. Applicable SimDis standard procedures for liquid vehicle fuels are ASTM D 7096-16 (gasoline, boiling point max. 280 °C); ASTM D 7213-15 (petroleum distillates, boiling point range 100 to 650 °C); ASTM D 7398-11 (FAME, boiling point range 100 to 650 °C).

One lever for adjusting the vapor pressure of gasoline is called RVP-trimming, which makes use of low-boiling paraffinic hydrocarbons produced during production of LPG/LNG. Among these, the following alkanes deserve particular mention (boiling point in parentheses):

n-butane (-0.5 °C); iso-butane (-11.7 °C);
n-pentane (+36.0 °C); iso-pentane (+28.0 °C); neo-pentane (+9.5 °C).

These light hydrocarbons belong to the condensate fraction (also termed "natural gasoline") recovered from well gas or shale gas processing and may be used as blendstock for ethanol fuel blends instead of common gasoline, as has been evaluated by Alleman et al. (2015).

The topic of vapor pressure adjustment, or more generally, the implication of widely differing boiling points among fuel blend constituents, is quite complex. As for example higher boiling components have higher viscosity than low-boilers, fuel pumps exert higher pressure gradients to provide sufficient amounts of liquid towards injectors. The lowest boiling components give rise to concerns on possible cavitation at the low-pressure (intake) side of fuel pumps, which would transfer gas bubbles into the fuel system causing vapor lock. According to this, most diesel engine manufacturers do not allow low-boiling additives like gasoline to be used to enhance the cold start performance of diesel engines.

Rapid ("flash") volatilization of low-boiling minor fuel components has also been used for special fuel formulations like water-diesel or ethanol-diesel blends; even blends of gasoline in diesel ("diesoline") have been investigated.

<u>Flammability, ignitability: flash point, auto-ignition point, octane and cetane rating</u>

Among properties associated with ignition behavior of combustible fluids, <u>flash point</u> relates to occupational safety in terms of fire hazard at regular handling (i.e. refilling and maintenance). Regulation of flash point by fuel standards only applies to diesel fuel due to its relatively low volatility, which consumers implicitly rely on with respect to safety precautions. Gasoline and liquid gases, in contrast, have flash points below ambient temperatures and thus pose higher risks for fire and explosions, which is reflected by higher safety class ratings.

The <u>auto-ignition</u> temperature of fuels is determined at atmospheric pressure and oxygen content, providing another safety parameter, but it is mainly relevant as a pre-estimate of ignitibility of vaporized, pressurized fuel facing conditions in an engine cylinder.

Generally, high-<u>cetane</u> number (CN) compounds are low-<u>octane</u> number (ON) compounds and vice versa, which can be roughly estimated from structural elements and has been correlated empirically; see for example the graphical compilation in Yanowitz et al. (2014), p. 13. This correlation is of more than theoretical importance because recent and possible future combustion strategies are likely to involve variable operation conditions adapted both from spark ignition (SI) and compression ignition (CI) modes, as for example with VW's CCS® combustion concept (Steiger et al. 2008; related to homogeneous charge compression ignition (HCCI) combustion, see Jeihouni et al. 2013). Consequently, tailor-made fuels may (or may even have to) combine constituents which are treated from a traditional viewpoint as *either* of gasoline *or* of diesel type. Janecek et al. (2016) recently published a paper discussing the correlation between CN an ON. Foong et al. (2014) published a list of measured octane numbers of various ethanol blends with gasoline and some surrogates ranging from E0 to E100.

<u>Chemical descriptors</u> to keep an eye on for all fuel types are water content, sulfur content, amount of unsaturated and aromatic compounds, and content of trace elements like metals, alkali and phosphorus. Specific fuel types like alcohols, biodiesel (FAME or HVO/HEFA), DME or OME, may come with typical minor constituents or impurities that potentially impair fuel combustion and functionality of vehicle parts.

Technical guidance documents that include supporting information have been published by manufacturers of engines, vehicles, and fuels. Such documents provide recommendations on fuel properties that in most cases closely follow mandatory fuel standards, but in some cases claim more stringent quality criteria in view of specific experiences on their fields of action. Well-known WWFC guidelines for diesel/gasoline, biodiesel and ethanol are to be mentioned in this context, as are the IMPCA (International Methanol Producers and Consumers Association) Methanol Reference Specifications dedicated to quality assurance of methanol fuels. Online documents are accessible via the following URLs:

> http://www.acea.be/uploads/publications/Worldwide_Fuel_Charter_5ed_2013.pdf
> http://www.acea.be/uploads/publications/20090423_B100_Guideline.pdf
> http://www.acea.be/uploads/publications/20090423_E100_Guideline.pdf

http://www.methanol.org/wp-content/uploads/2016/07/IMPCA-Ref-Spec-08-December-2015.pdf

Pre-sale actors like operators of storage facilities and pipeline grids must perform stringent quality control on fuels before they are allowed to enter their infrastructure, especially if containments and auxiliaries are charged with variable types of fluid or gaseous media that each come with special trace components and reactive potentials. An example of corresponding specification requirements is provided by Buckeye Partners L.P. within a shipping information handbook (fungible/terminal product grade specification, 2016), which is available online at

http://www.buckeye.com/LinkClick.aspx?fileticket=9iaDRdk9OOs%3D&tabid=125 (pipelines)

and

http://www.buckeye.com/LinkClick.aspx?fileticket=8EMhbgilmTs%3d&tabid=125 (terminals).

Furthermore, it must be noted that existing fuel standards refer to main fuel types of relevance for public markets, i.e. gasoline, diesel, biodiesel (FAME or HVO/HEFA), ethanol, and LNG/LPG, and that some of the emerging advanced fuels do not (and cannot) fulfill all of the respective specifications, due to their inherent physicochemical properties. Failure to meet a given specification does not necessarily mean a new fuel type is unsuitable for engine use, but it may imply that its properties will require changes to materials of construction and machinery, or that use via blending may be needed.

5 Liquid advanced biofuels for road transport

There are many different liquid biofuels that can be applied as blends or substitutes to conventional diesel or gasoline fuels under investigation by various international RD&D activities. The following advanced liquid biofuels are considered here since they best match commonly accepted criteria, as documented and implemented in different national mobility strategies; e.g., see IEA-AMF 2014, ABFA[1] or ACEA[2].

- Hydrotreated vegetable oils or esters and fatty acids (HVO/HEFA)

- Synthetic biobased fuels such as biomass-to-liquids (BTL, Fischer-Tropsch fuels (FT)), methanol, dimethyl ether (DME) and oxymethylene dimethyl ether (OME)

- Lignocellulosic ethanol

- Liquefied biomethane (Bio-LNG)

For each of these biofuels facts sheet tables have been elaborated addressing the following issues[3]:

- Typical feedstock representative of the bio-based resources in use or being researched today

- Typical fuel production process characteristics identifying the most important technology issues and providing information on typical or potential coproducts of the process

- Examples of main technology providers for overall plant design, construction and operation

- Technology and fuel readiness level (respectively TRL and FRL), with TRL defined according to the European Commission, which outlines in detail the different research and deployment steps (1 – basic principles observed, 2 – technology concept formulated, 3 – experimental proof of concept, 4 – technology validation in lab, 5 – technology validation in relevant environment, 6 – demonstration in relevant environment, 7 – demonstration in operational environment, 8 – system completed and qualified, 9 – successful mission operations) (EC 2010); and FRL defined according to CAAFI: R&D leading FRL 1-5, certification FRL 6-7, and business & economics FRL 8-9 (CAAFI 2010)

- Typical fuel production costs (normalized to 2015 prices) as an indicator of economic competitiveness including the caveat that cost estimates are highly dependent on accurately assessing the respective TRL/FRL as well as the rigor in plant design, and can also depend on the regional framework and available infrastructure; and also on the methodology used for cost calculation.

- Typical GHG emissions for well-to-tank (WTT) fuel production and distribution as an indicator for competitiveness for climate friendliness including the notice that life cycle assessment (LCA) results are also highly depending on the same factors as fuel production costs and well as the methodology for their calculation (e.g. EU RED or US RFS).

---

[1] ABFA (Advanced Biofuels Association) is focused on leading America's green economy, where its member companies have the ingenuity and entrepreneurship to fuel a sustainable energy future.
[2] ACEA (European Automobile Manufacturers' Association) represents the 15 Europe-based car, van, truck and bus manufacturers: BMW Group, DAF Trucks, Daimler, Fiat Chrysler Automobiles, Ford of Europe, Hyundai Motor Europe, Iveco, Jaguar Land Rover, Opel Group, PSA Group, Renault Group, Toyota Motor Europe, Volkswagen Group, Volvo Cars, and Volvo Group. More information can be found on www.acea.be.
[3] With no claim to completeness.

- (very brief) SWOT with strengths (S), weaknesses (W), opportunities (O) and threats (T) compared to conventional biofuels like FAME or HVO/HEFA biodiesel, ethanol production from sugar or starch feedstock, and in terms of their general applicability for use in transport.

Additionally, this chapter summarizes fuel properties and emission characteristics of these fuels (see Table 25). In addition, new aspects of FAME biodiesel are provided.

5.1   Hydrotreated vegetable oils and hydroprocessed esters and fatty acids (HVO/HEFA)

Hydroprocessing (also called hydrotreating, or occasionally hydrogenating) can be applied to fats and oils from crop or animal origin to produce HVO/HEFA type biofuels. HVO and HEFA are highly saturated, branched and straight-chain hydrocarbons without oxygen (or other heteroatoms), produced from oleaginous feedstock via a succession of decarbonylation, decarboxylation and hydro-deoxygenation processing steps. These fuels exhibit comparably better fuel properties than FAME biodiesel however they compete for same types of oleaginous feedstock. Worldwide, production has increased forty-fold, from about 0.1 million tons produced in 2008 to more than 4 million tons in 2016 (Naumann 2016). Table 17 summarizes information on HVO/HEFA biofuels.

HVO/HEFA biodiesel fuels represent one important type of synthetic, highly upgraded drop-in fuel that is fully compatible with existing infrastructure and vehicle design. Their heating value closely matches that of fossil diesel and they also enable lower-sooting emission operation of vehicle engines. Some trade-off exists because of the paraffinic nature of HVO/HEFA, which causes low lubricity and viscosity and high cetane number. These deviating properties (especially low lubricity) have to be corrected by way of blending (see e.g. Lapuerta et al. 2011).

Emissions

HVO was tested for emissions as both neat fuel and as a blend with diesel fuel. The results of seven test series are summarized in Figure 2. On average, the emissions of hydrocarbons and carbon monoxide are reduced by 30 to 40% using HVO rather than fossil diesel. Also the emissions of particulate matter decreased by 25% on average. Against the trend of the other emissions, nitrogen oxide emissions show no clear trend. The emissions of passenger cars (green dots) are in general increased, while the emissions of heavy duty vehicles tended to be lower.

Additionally, Happonen et al. (2013) investigated exhaust emissions and emitted soot particle hygroscopicity using 80:20 mixtures of HVO and dipentylether. The authors detected general improvements in emission quality.

Table 17: Fact sheet HVO/HEFA (data base DBFZ 2016)

HVO/HEFA	Short description
Typical feedstock	Usually vegetable oils and fats (mainly palm, used cooking oil; for aviation jatropha, camelina); alternatively tall oil and (often under R&D&D) algae oils, tobacco, but also pyrolysis and hydrothermal based oils
Typical process characteristics	Multistep hydrotreating of feedstock with hydrogen (usually out of natural gas steam reforming or internal naphtha reforming), comparable to conventional refinery processes Multiproduct plant: HVO/HEFA diesel or jet fuel, naphtha, propane/butane (and if directly with annex oil mill in case of rape or soya also extraction meal as fodder)
Main technology provider (examples)	Single plant capacities about 0.1 to 0.8 million t a^{-1}, regional focus EU, US, Indonesia Neste Cooperation (total capacities: 1.9 million t a^{-1}), UOP/Eni (0.55 million t a^{-1}), UOP/Galp Energia + Petrobras (0.22 million t a^{-1}), Axens/TOTAL (0.5 million t a^{-1} startup 2017), UPM (0.1 million t a^{-1}), Preem/Sunpine (0.1 million t a^{-1}), UOP/Diamond Green Diesel (0.27 million t a^{-1} startup 2017), Dynamic fuels (0.2 million t a^{-1})
TRL/FRL	9/9, currently clear focus on diesel fuel production; ASTM certified jet fuel as HEFA-SPK (only in batches as requested by costumers), ASTM certification for HEFA Diesel+ expected for 2017
Typical fuel production costs	19 to 47 USD per GJ Mainly feedstock price, specific TCI about 360 to 495 USD per kW fuel
Typical GHG	5 to 76 kg CO_{2eq} GJ^{-1} (according RED)
S | Strength	Higher fuel quality compared to FAME. Can be blended up to about 100% v v^{-1} (but usually limited by 20% to 30% v v^{-1} due to density given by fuel standards); high ignition quality (cetane number) and no aromatic content enable low-emission combustion; material compatibility, storage stability and calorific value comparable to conventional diesel fuel
W | Weaknesses	Competitor with biodiesel/FAME regarding vegetable oils and used cooking oils or animal fats as feedstock and therewith part of the debate on feedstock availability and impact on d/iLUC; especially in context of palm oil use; low density disabled high blending rate given by fuel standards
O | Opportunities	Increasing installed capacities worldwide, change to so called advanced feedstock (like algae-based oils, bio-oils out of pyrolysis or hydrothermal processes) possible
T | Threats	Challenge of (renewable) hydrogen availability for hydrotreating processes

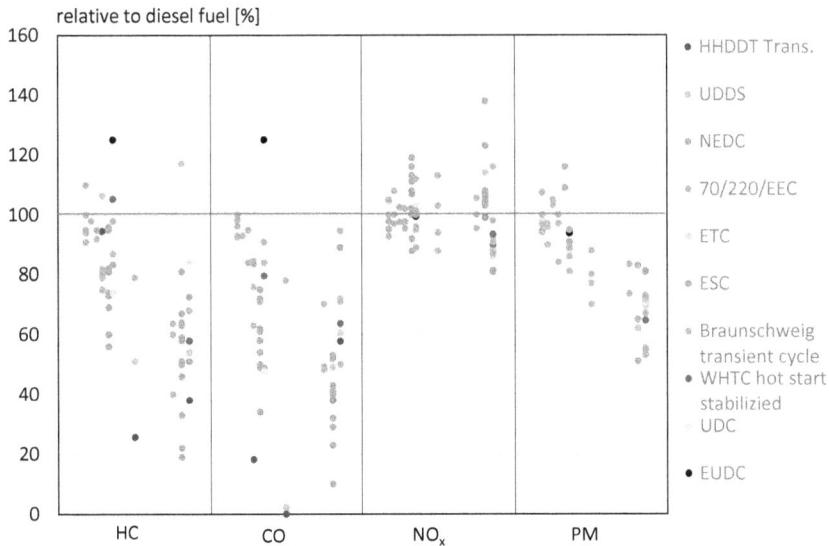

relative to diesel fuel [%]

Legend:
- HHDDT Trans.
- UDDS
- NEDC
- 70/220/EEC
- ETC
- ESC
- Braunschweig transient cycle
- WHTC hot start stabilizied UDC
- EUDC

X-axis categories: HC, CO, NO$_x$, PM

Figure 2: Overview for regulated emissions of HVO and HVO blends relative to diesel fuel. The blend proportion starts with 10% left side and goes up to 100% right side of each emission component. The single tests are coloured by the test cycles used. (Erkkil\am et al. 2011, Pabst 2014, Kuronen et al. 2007, Singer et al. 2015, Prokopowicz et al. 2015, Karavalakis et al. 2016, Rantanen et al. 2005)

5.2 Biomass-to-liquids fuels (BTL)

The production of synthetic biomass-to-liquids (BTL) is basically characterized by three main steps after appropriate biomass pre-treatment: (i) gasification of lignocellulosic or pre-treated biomass to produce a raw gas, (ii) treatment of raw gas to produce synthesis gas (syngas), (iii) catalytic synthesis of syngas to produce synthetic biofuels, and (iv) final treatment to meet product specifications. Despite a long history of development and a broad variety of system configurations tested to date, no commercial market break through has yet been realized for producing FT synfuels via biomass gasification. The most common approach under RD&D is FT synthesis, but alcohols like methanol and ethers like DME and OME, whose properties are discussed separately here, are sometimes included in the group of BTL fuels. The first production plants producing methanol from biomass via syngas are running. This methanol is also converted to ethanol and sold as cellulosic ethanol (Enerkem, 2017).

FT processes can use a diverse range of carbonaceous feedstock, even fossil starting materials like coal (CTL), refinery gas (fossil GTL), or eventually feed gases produced by means of renewable energy (power-to-liquid, PTL), FT or XTL. At present, descriptive summarizing terms for FT-type processes do not discern between low- and high carbon footprint products. The potential of FT fuels for sustainable mobility only arises when processing renewable or waste biomass feedstock, thus furnishing products for which BTL is a legitimate term.

FT fuels are mainly paraffinic in nature and thus closely resemble HVO/HEFA from chemical and physical property points of view, including their low lubricity and viscosity. A succession of hydrocarbon buildup steps during FT synthesis allows the relative amounts of different chain length types to be modified, thereby offering a potential FT product palette that can span from meeting gasoline to diesel fuel specifications. Low-soot combustion can be achieved by formulating FT blends as fully compatible premium fuels. Performance and emission testing results are still relatively scarce to come by due to relatively high production costs and thus limited commercial usage to date.

Table 18 provides summary information on FT fuels.

Current pilot/demonstration plants are often run primarily in shorter test campaigns. Therefore, production of significant volumes of FT fuels from biomass should not be expected in the short- or medium-term.

Emissions

Emission data from BTL fuels are quite limited. Most of the BTL is produced from biogenic synthesis gas via FT synthesis. However, most of research on emissions of FT fuels was done with GTL, for which the final fuel product was proposed to have the same property requirements as fuels produced via BTL. Nevertheless, two articles report emissions of BTL versus diesel fuel (Swain et al. 2011, Ogunkoya and Fang 2015). Additionally, BTL can also be produced from crude tall oil (CTO) via hydrogenation. Neimi at al. (2016) describe the emissions of CTO-based BTL fuel, and its relative emissions are summarized in **Error! Reference source not found.**.

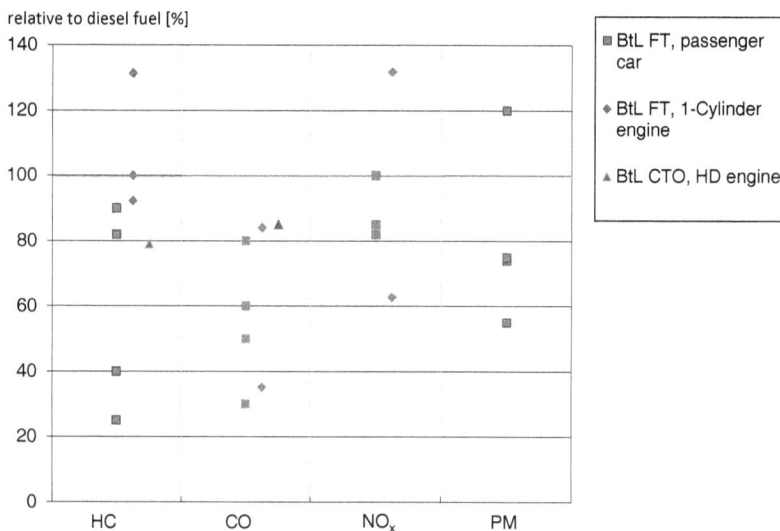

Figure 3: Overview for regulated emissions of BTL fuels relative to diesel fuel. Investigations were carried out on a passenger car, a 1-cylinder engine, and a heavy-duty (HD) engine. (Swain et al. 2011, Ogunkoya und Fang 2015, Niemi et al. 2016)

Table 18: Fact sheet Fischer-Tropsch fuels (based on DBFZ database 2016 as well as Perimenis 2010, Mueller-Langer 2014, IFP 2016, Rauch 2016)

FT fuels	Short description
Typical feedstock	Wood (industrial wood, waste wood, short rotation coppice), stalk material (mainly straw, triticale whole plants, miscanthus), wood-based black liquor, US: also municipal waste
Typical process	Mechanical treatment (e.g. grinding, crushing), thermal pretreatment (e.g. pyrolysis, drying, smoldering), gasification, gas purification, gas treatment (e.g. scrubber, filter, absorption, reforming, shift-reaction), FT synthesis, product treatment (e.g. hydrocracking, distillation, isomerization) Multiproduct plant: BTL diesel or jet fuel, waxes, naphtha, heat and power
Main technology provider (examples)	Regional focus EU, US BioTfueL demonstrator (Avril, Axens, Total, Thyssen Krupp, IFP and others with 3 t h^{-1} feedstock; IFP 2016), Fulcrum Bioenergy/Tesoro (0.2 million t a^{-1} feedstock), Red Rock Biofuel/Velocys; small-scale: e.g. Velocys, Ineratec, Bioenergy2020+/TU Vienna
TRL/FRL	5 to 6/5
Typical production costs	18 to 62 USD per GJ mainly capital investment and feedstock price, specific TCI about 2,600 to 4,260 USD per kW fuel
Typical GHG	7 to 100 kg CO_{2eq} GJ^{-1} (according RED)
S | Strength	Higher fuel quality compared to FAME. Can be blended up to about 100% v v^{-1} (but usually limited by 20% to 30% v v^{-1} due to density given by fuel standards like EN 15940); high ignition quality (cetane number) and no aromatic content enable low-emission combustion; material compatibility, storage stability and calorific value comparable to conventional diesel fuel
W | Weaknesses	Comparably low overall efficiency and the production of a wide range of different aliphatic hydrocarbons which makes intensive product separation and treatment necessary for the production of applicable fuels; therefor comparably cost intensive; low density disabled high blending rate given by fuel standards
O | Opportunities	Handling and application of FT fuels internationally known from GTL, FT fuel still of interest for different road and aviation applications (production of bio jet fuel that already is ASTM certified)
T | Threats	Ongoing delay in commercialization along innovative chain for FT fuels, challenge to be price competitive with HVO/HEFA which shows similar fuel properties

To get a better impression of the emission characteristic of FT fuels, Figure 4 shows an overview of published emission data of GTL-based FT fuels, which clearly illustrates that all pollutant emissions categories are reduced using FT-fuel.

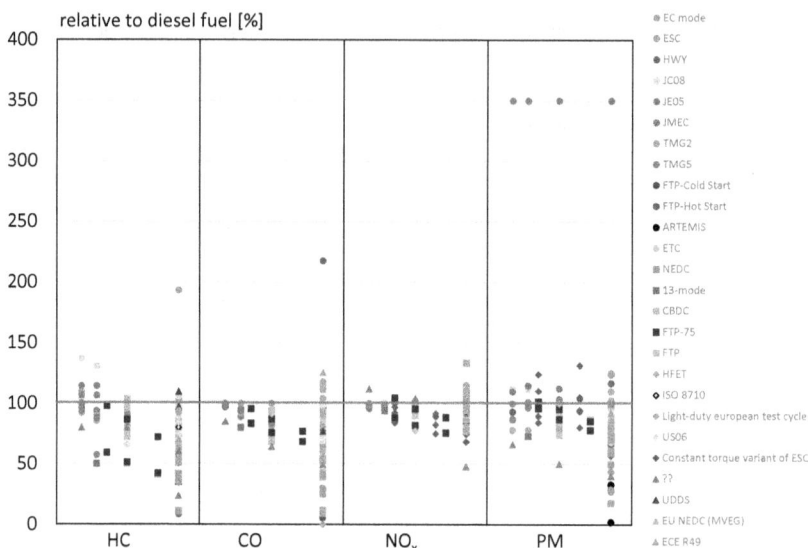

Figure 4: Overview for regulated emissions of GTL and GTL blends relative to diesel fuel. The blend proportion starts with 5% left side and goes up to 100% right side of each emission component. The single tests are coloured by the test cycles used. (Original literature is not cited in this review.)

5.3   Dimethyl ether (DME)

Fossil-based dimethyl ether (DME) has been used traditionally as an energy source in countries including China, Japan, Korea, Egypt and Brazil. DME is produced in considerable amounts as an industrial chemical by catalytic conversion of methanol with early applications as a spray propellant. DME has also received attention as an alternative fuel but alternatively it can be used as a process intermediate in catalytic production of high-octane synthetic gasolines (dimethyl ether-to-fuels pathway, see research note by NREL, 2016). A recent *status quo* report on DME issues is provided by McKone et al.'s DME Tier I Report (2015). Ongoing RD&D considers renewable methanol production and dehydration processes combined within one reactor, such that DME is produced directly from synthesis gas slightly more efficiently than methanol. Azizi et al. (2014) review DME production technologies. DME's fuel properties are similar to LPG thus allowing the use of the same infrastructure. Table 19 provides summary information on DME fuel.

DME is the simplest aliphatic ether, its molecules consisting of two CH_3-groups connected by a bridging oxygen atom. While gaseous at ambient conditions, it is easy to condense and therefore can be handled similar to LPG or LNG. As it is a single chemical compound, it displays a distinct boiling point (-24 °C) instead of a boiling point range and can be provided in a highly pure quality. Trace impurity concentrations to be controlled to meet fuel specifications are methanol, methylformate, CO_2/CO and water from production, as well as evaporation residues similar to fossil fuel gases. DME exhibits low lubricity, making blend formulations with LNG/LPG hydrocarbons a potentially viable option.

Moreover, fueling with DME is not confined to gas engines but rather offers promising applications in combination with liquid fuels, catalytic on-board generation and also potential use as make-up component in exhaust gas to support catalytic NO_x reduction. DME is also used as a fuel in CI-engines because of its high cetane number, however the injection system must be modified to use DME.

Table 19: Fact sheet DME (based on DBFZ data base 2016 as well as Burger 2010, Sakar 2011, Fleisch 2012, Silalertruksa 2013, Zeman 2013, Tuná 2014, bioliq® 2016)

DME	Short description
Typical feedstock	Wood based black liquor, wood, solid organic waste (wood, municipal solid organic waste, straw), biogas
Typical process	Mechanical pretreatment (crushing), fast pyrolysis (in case of decentral collecting for better transportation), drying, gasification, gas purification, gas treatment (reforming, water gas shift, Sulphur removal, CO_2/water-removal), indirect synthesis: methanol synthesis + methanol dehydration, direct synthesis: DME synthesis + methanol as byproduct, product purification (distillation)
Main technology provider	Regional focus: EU, US Chemrec/LTU (black liquor – pilot plant/out of operation), KIT (solid organic waste – pilot plant), OberonFuels (biogas/natural gas – pilot plant, offers commercial small scale units)
TRL/FRL	4-6/5
Typical production costs	16 to 30 USD per GJ
Typical GHG	1 to 72 kg CO_{2eq} GJ^{-1} (according RED)
S | Strength	Low soot emission, established synthesis technology
W | Weaknesses	Low energy density (around 50% of fossil diesel); gaseous fuel at normal temperature and pressure; low viscosity and poor lubricity; material compatibility of fuel-carrying components not assured
O | Opportunities	LPG replacement (LPG/DME blends already used in different regionals like in Asia) also alternative fuel for diesel engines
T | Threats	Development of OME2-5 fuel (methanol/DME derived, also low soot emission, diesel like handling)

Emissions

It is useful to review earlier work on DME fuel use. The standard configuration for neat DME comes with a pressurized vehicle fuel tank as in LPG/LNG vehicles, but it is possible to apply liquid fuel-DME blends using a moderately pressurized diesel tank with modification of the injection system. Spray atomization in such applications is boosted by the flash-boiling effect of low boiling DME. Due to the adaption of the engines to the fuel, a direct comparison of the emissions, like for HVO, is not possible. Emissions trends of DME fuels are presented in the following paragraphs:

Hewu & Longbao (2003) worked with diesel fuel-DME 90:10 blends and documented clear emission benefits for smoke, NO_x and HC. Ying et al. (2008) performed a study on diesel fuel-DME 90:10, 85:15

and 80:20 blends including solubility and spray behavior on a commercial four-cylinder DI diesel and noted emission benefits for PM and NO_X, while HC and CO showed less favorable values. Biodiesel-DME blends in the range 0:100 to 100:0 were evaluated by Wang et al. (2011), who observed general decreases of criteria emissions, though unambiguously only for smoke and NO_X, while CO emission decreases were only evident at high loads and HC emission increases at low loads. Important to note, "smoke reduction" has to be inspected more precisely, as a shift to smaller particle sizes may occur requiring a more differentiated assessment of particle emissions.

Youn et al. (2011) reported results for a four-cylinder diesel engine alternating operation on diesel fuel and DME. Injection timing was varied and spray characteristics were documented. Taken from three speed settings (1000/1500/2000 rpm) at constant 20% load, soot emissions for DME were almost negligible, HC and CO emissions showed clear benefits, but NO_X displayed slight disadvantages irrespective of injection timing.

Reviews on DME application in diesel engines are provided by Park and Lee (2013, 2014). A comprehensive review on DME application in diesel engines with respect to upcoming emission standards is provides by Thomas et al. (2014). The pressure and injection behavior of DME-biodiesel mixtures have been studied on an injection test stand (no engine) by Hou et al. (2014).

DME fumigation was investigated by Armbruster et al. (2003). DME is catalytically generated *in situ* from methanol and applied as fumigation component in a methanol vehicle adapted from commercial diesel engine. Specific emission testing was not performed.

Generally, comparing performance and emissions of neat diesel and neat DME engines implies comparing two different engines, since equipment modifications according to different fuel properties have to be made. Examples are investigations by Zhang et al. (2008), Fang et al., (2011), Cha et al. (2012) and Szybist et al. (2014, baselines from comparison of common diesel and specially fitted DME-fueled prototype truck). Beneficial effects regarding particulates and often NO_X are claimed, and mixed outcomes regarding gaseous criteria emissions, depending on driving conditions; high exhaust gas recirculation (EGR) rates are tolerated. Gill et al. (2014) investigated injector nozzle shaping and other factors of combustion control in a single-cylinder DME engine and obtained additional improvements for criteria emissions.

Zhao et al. (2014) described an engine with premixed DME air intake (up to 40% DME) and conventional diesel fuel injection. They found a benefit in PM emissions but an increase of NO_X emissions at standard operation mode. This allowed an increase in the EGR rate which resulted in less NO_X emissions at unchanged PM emissions.

Park et al. (2011) and Choi et al. (2013) studied DME as a make-up component in exhaust gas of a DME engine. DME is reformed via separate catalyst to liberate H_2 into exhaust gas, which promotes NO_X reduction in lean NO_X trap exhaust treatment. DME dosage was as low as 1% and NO_X turnover was improved up to 20% depending on temperature and catalyst; benefits for NO_X conversion can be expected to grow for specific catalyst optimization.

The similarity in handling of DME and LPG recommends dual-fuel applications (both in CI and SI engines). Konno and Chen (2005) studied the performance of methane-DME fuel mixtures on HCCI engine performance and found high ignition improvement by DME. Jamsran and Lim (2014) applied DME-n-butane blends to an one-cylinder HCCI engine and found an optimum emission decrease at a DME-butane ratio of 1:0.6. Flekiewicz et al. (2014) investigated SI engine characteristics with both hydrogen enriched natural gas and an LPG-DME blend and found slight criteria reductions for higher

H₂ share in LNG and some criteria increases for higher DME share in LPG. Lee et al. (2009) operated a SI engine with DME blended LPG fuel and stated that "hydrocarbon and NO_x emissions were slightly increased when using the blended fuel at low engine speeds", and depicted 1800 rpm diagrams for NO_x, HC and CO overall do not show emission improvements for 10, 20, or 30% DME-LPG blends. Note that this is an SI application, and that the low octane rating of DME limits higher levels of blending. Again Lee et al. (2011a) operated an SI engine on LPG (major component n-butane) modified with propane or DME, finding propane better suited than butane for DME blending, as higher ON of propane preferred over lower ON of DME. FTP-75 cycle tests revealed similar emission results for LPG and an LPG-20% DME mixture. Another investigation by Lee et al. (2011b) employing a single-cylinder diesel engine revealed that blending of n-butane into DME as main fuel component (DME-butane 90:10, 80:20, 70:30, 60:40) did not improve gaseous criteria emissions.

To summarize, DME potential for exhaust particulate reduction remains under investigation, but gaseous criteria emissions do not decrease in all cases to a substantial degree, especially in SI applications. Further engineering efforts are required to fully exploit DME potentials.

Further information of DME topics

Basic thermodynamic properties of DME from state equation calculations can be found in Teng et al. (2004) and Zhao et al. (2005).

Vapor pressure and lubricity of DME-diesel oil blends are described in Li et al. (2007).

A 2008 review by Arcoumanis et al. (2008) provides information on DME supply chains, sustainability and engine performance issues being studied at that time, with many general conclusions still valid.

Insight into pre-standardization work on DME fuels is provided in a presentation by Oguma et al. made at the 7th Asian DME Conference 2011; see URL:

> http://aboutdme.org/aboutdme/files/cclibraryfiles/filename/000000001958/7asiandme_aist_oguma.
> pdf

5.4   Oxymethylene dimethyl ether (OME)

Oxymethylene dimethyl ether (OME) of the type $CH_3(OCH_2)_nOCH_3$ (n = 1-6) is gaining increasing interest as a diesel oxygenate. While OME can be produced from methanol via different routes, sustainable OME production from biomass or renewables on technical scale is still a challenge. So far, only OME1 is produced in commercial quantities (PubChem, 2017). Notation of OME-homologues (most commonly OME1 up to OME6) in the literature is somewhat variable from author to author; keyword search should take into account: POMDE (poly(oxymethylene) dimethyl ether), POMDME (poly(oxymethylene) dimethyl ether), POMDAE (poly(oxymethylene) dialkyl ether), PODE (poly(oxymethylene) dimethyl ether); for OME1: DMM (dimethoxymethane) or methylal.

Synthesis conditions as well as product compositions and purities show considerable scatter, which in turn impedes correlation and interpretation of emission testing results. As for DME, which is structurally related (sometimes referred to as "OME0[zero]") and likewise formed via catalytically mediated reactions, methyl formate is an impurity of concern, as are products of hydrolysis corresponding to back-reaction. Synthesis conditions also determine the relative amounts of lower/higher homologues.

The most convenient as engine fuels are the homologues OME1 up to OME6 with boiling points ranging from about 42 to 280 °C, and high cetane numbers from OME2 upwards. The fuels have been investigated intensely by Lumpp et al. (2011), Härtl et al. (2014), Feiling et al. (2016) and Härtl et al. (2017). All of these OME homologues allow engine fueling with almost soot-free emissions, though such performance might be compromised by impurities or poorly specified homologue distributions. Available quantities of adequate OME fuel formulations are still limited which has hampered broad engine testing. According to boiling range and ignition behavior, CI applications appear preferable; however, CI engine operation with neat OME fuels would still face some limitations or challenges resulting from higher density and lower viscosity as well as lubricity mismatching diesel standards. Table 20 summarizes information on OME.

Publications dealing with production and properties of OME-type fuels are from Burger et al. (2010) and Lautenschütz et al. (2016), Lautenschütz et al. (2017), Deutsch et al. (2017) and Oestreich et al. (2018). Investigations on synthesis conditions and improvements in OME product quality are described in publications by Zhao et al. (2011b), Zhang et al. (2014), Zheng, Tang et al. (2015a) and Ouda et al. (2017). A comprehensive investigation on OME production pathways and catalysis conditions has been consolidated in the Ph.D. work by Lautenschütz (in German) in 2015. In this context, Goncalves et al. (2017) have carried out a computational study for a better understanding of the underlying reaction mechanisms.

Schmitz et al. (2016) investigated the production costs for OME fuels. They found that the price would be between 400 and 900 USD per ton depending on facility scale and the methanol price. Therefore, OME production is economically competitive with conventional diesel fuel production. In a very recent study, OME production costs have been estimated considering different types of biomass feedstock. According to this study, production costs are in the range from 1.66 to 1.93 USD per liter for 500 tons per day of dry biomass (Oyedun et al. (2018)).

Li et al. (2017) tested the effect of OME addition to diesel fuel on spray and atomization characteristics, while OME properties and results from engine emission testing have been described by Lumpp et al. (2011), Härtl et al. (2014), Feiling et al. (2016) and Härtl et al. (2017) in the publications mentioned above.

Table 20: Fact sheet OME (based on DBFZ data base 2016 as well as Burger et al. (2010), Maus et al. (2014), Zhang et al. (2014), Sauer et al. (2016), Schmitz et al. (2016) and Zhang et al. (2016))

OME	Short description
Typical feedstock	Wood, solid organic waste (wood, municipal solid organic waste, straw), biogas
Typical process	Thermo-chemical pathway: Mechanical pretreatment (crushing), drying, gasification, gas purification, gas treatment (reforming, water gas shift, Sulphur removal, CO_2/water-removal, methanol synthesis, intermediate synthesis from methanol (formaldehyde, methylal, trioxane), OME synthesis, purification (distillation)
Main technology provider	Several plants in China for conventional OME with commercial quantities Research activities esp. in the EU, e.g. by KIT (OME via OME1/trioxane see: Lautenschütz et al. (2017), OME via methanol/formaldehyde, see: Oestreich et al. (2017) and OME via DME/trioxane, see: Haltenort et al. (2018)); patents mainly hold by BP, BASF, Chinese companies
TRL/FRL	3 to 4/3
Typical production costs	33 to 50 USD per GJ (for non-renewable OME) mainly capital investment, methanol price
Typical GHG	No data available
S | Strength	OME3-5: efficient NO_x reduction, diesel like handling and properties; high oxygen content reduces soot emissions; high ignition quality, reduction of engine noise; highly efficient application in internal combustion engines
W | Weaknesses	Low calorific value; material compatibility of fuel-carrying components needs to be adapted; OME1/ OME2: only with additives usable (e.g. OME1 low cetane number and viscosity); low boiling temperature; OME3-5: high density (> 1.000 kg m^{-3})
O | Opportunities	Alternative fuel for diesel engines (up to 100%)
T | Threats	Processes for renewable products are not state of the art; OME in strong competition to other diesel drop in fuels; formaldehyde production needs to be improved

Emissions

Though not numerous, several promising studies on OME engine performance and emission behavior have been published. Already mentioned are short articles in MTZ worldwide edition (in English): Lumpp et al. (2011), Härtl et al. (2014), Feiling et al. (2016) and Härtl et al. (2017). Note that viscosity and lubricity of OME do not fully comply with fuel standards, so OME fuels for engine testing often are additivated, which has to be considered prior to data comparisons.

Oxygenate screening and neat OME1 testing in heavy-duty single-cylinder diesel are reported by Härtl et al. (2015). OME1 plus 13 more oxygenates (diverse glycol type esters, each blend adjusted to same oxygen level), have been emission screened at 1200 rpm/medium load in comparison to ULSD reference (fuel switching). All oxygenates provided soot reduction, with performance depending on C/O- and H/C-ratio, plus structural factors. The highest soot reduction was achieved by OME1, which was subjected to a test series with variable air/fuel and EGR ratios (additivated "OME1a" for sufficient viscosity, lubricity and CN). Neat OME1a, 25:75 OME1a-diesel and 95:5 OME1a-diesel blends have been tested, and even these latter low diesel amounts increased sooting such that a soot-NO_x tradeoff occurs. Neat OME provides extremely low soot at any EGR rate without soot-NO_x tradeoff, and no rise in particle number (PN). CO and HC emissions rise after diesel oxidation catalyst (DOC) at high EGR ratios with a lambda value of one and lower (insufficient oxygen content in exhaust gas). Formaldehyde was not detectable, but methane rose significantly at low equivalence ratio due to ineffective DOC at high CH_3 radical inventory in exhaust gas. In a very recent study, the ignition behavior of OME1-gasoline blends has been investigated, due to the comparatively low cetane number of 28 in the case of OME1. Thus, it could be demonstrated that the use of OME1 is not restricted to diesel engines.

Iannuzzi et al. (2016) generated data on emission performance of OME1, OME2 and an OME2-4 mixture as compared to common diesel from constant volume chamber experiments with optical access. Neat OME fuels proved to give almost sootless emissions with nonlinear soot reduction in OME2-diesel 5%, 30%, and 50% blends. Particle sizes of soot decreased markedly.

Heavy-duty diesel engine application of two OME-diesel blends is reported by Liu et al. (2016). Mixtures of higher homologues OME3-6 as a blend component in ratio diesel-OME ratios of 85:15 and diesel-OME 75:25 have been tested. Results from three engine load settings show clear OME benefits for CO only at high load and high EGR ratios, HC benefits only at low- to medium load, soot benefits for all loads and EGR rates and NO_x benefits for low- and medium load. World harmonized stationary test cycle (WHSC) results of weighted criteria emissions upon EGR adjustments to control NO_x show improved soot-NO_x tradeoff and reduction potentials for HC and CO by adding OME.

A recent investigation on OME HCCI engine combustion has been published by Wang et al. (2016) using an OME2-4 fuel in which OME3 was the major component. Almost soot free and low NO_x emissions were observed, but HC and CO emissions performance was less satisfactory. Sun et al. (2017) performed a flame reaction kinetics study on OME3 employing a laminar burning test apparatus, constant volume combustion chamber, molecular species mass spectrometry and modeling. Results for species detection and formation rates contribute to a better understanding of fuel turnover pathways and characteristics.

The emission data of Wang et al. (2015), Liu et al. (2017), Liu et al. (2016), Feng et al. (2016), Lumpp et al. (2011) and Härtl et al. (2015) are summarized in Figure 5. This figure shows clearly the possibility of OME fuel blends to reduce HC, CO and PM emissions.

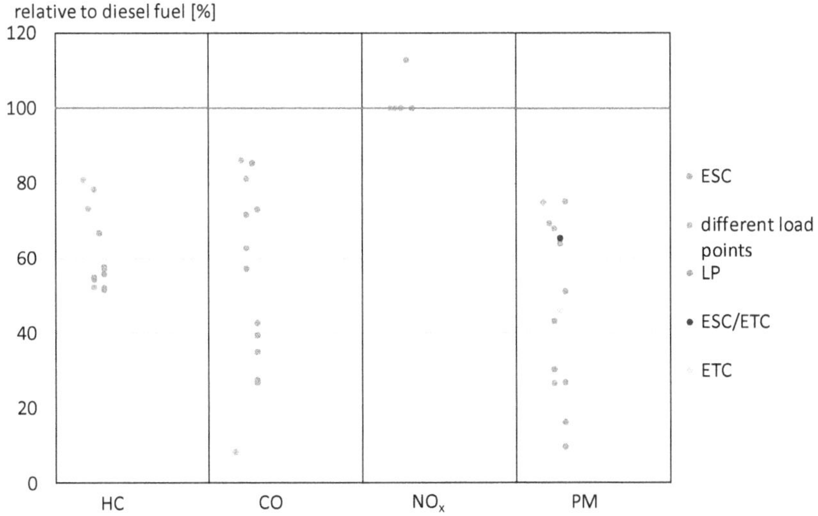

relative to diesel fuel [%]

Figure 5: Overview for regulated emissions of OME$_{3-5}$ blends (5, 10, 15, 20, and 25% from left to right in each column) relative to diesel fuel. The single tests are coloured by the test cycles used. (Wang et al. 2015, Liu et al. 2017, Liu et al. 2016, Feng et al. 2016, Lumpp et al. 2011 and Härtl et al. 2015).

5.5 Alcohols as fuels

Methanol and ethanol, both simple aliphatic alcohols, have been used as fuels since earliest times of engine development. They are produced on large-scale by conversion of biomass and/or gas-to-liquid processing. Benefits with respect to improved exhaust gas quality, especially reduced particulate emissions, have been observed both from blending with gasoline and diesel. On the other hand, as moderately polar solvents, they are completely miscible with water which in turn means high susceptibility to water uptake. Exceedingly high water content of hydrocarbon-alcohol blends causes phase separation unless emulsifiers are added; furthermore, susceptibility to oxidative processes and corrosion will grow and exhaust gas quality can be affected.

Recent and comprehensive articles dealing with methanol and ethanol fuel issues are provided by Ghadikolaei (2016), Yusri et al. (2017), and Geng et al. (2017). Quality surveys on ethanol fuels are given by Alleman (2013) and Cummings (2011). Ternary systems comprising gasoline, methanol and ethanol (GEM) or even their hydrous blends (formally quaternary blends) have been characterized and adapted to yield iso-stoichiometry with respect to drop-in performance, serving the needs of a large fleet of flex-fuel vehicles (Turner et al. (2013), Pearson et al. (2014), and Sileghem et al. (2014)).

Higher aliphatic alcohols with appropriate boiling properties are suitable as neat fuels or blend components both for SI and CI application. Both fossil- and biomass-based production pathways exist, and production cost remains a main factor hindering large-scale use of higher alcohols as engine fuel. For the butanol to decanol range, their generally low cetane and high octane numbers favor SI applications; nevertheless, diesel blends containing butanol or octanol have been formulated according to fuel standard target values and tested successfully in terms of fuel system compatibility and exhaust emissions (Schaper et al. 2017). The concept is working well if blending allows balancing of favorable and unfavorable properties to match a desired parameter value. Additional methods of achieving fuel property benefits are introduced this way; in the case of higher alcohols, they can act as oxygen donators, solubilization aids or to adjust bulk physical properties like viscosity and vapor pressure.

For a general overview on supply and performance of gasoline aliphatic oxygenates, see Aakko-Saksa et al. (2011). A comprehensive study on designing tailor-made biofuels is provided by Dahmen and Marquardt (2016). Physical-chemical properties of several oxygenate blends (methanol to dimethylfuran, a cyclic ether) in conjunction with specification needs are described by Christensen et al (2011).

5.5.1 Methanol

Like for FT fuels, for methanol produced from biomass there are just few pilot/demonstration plants (e.g., Enerkem in Canada based on municipal waste and BioMCN in the Netherlands based on biogas reforming). Zhen and Wang (2015) published an overview of production processes. The majority of methanol is produced from natural gas (methane) and coal. Methanol is used as base chemical for the production of formaldehyde, acetic acids, and other chemicals. In the energy/fuel sector methanol is used for the production of MTBE, DME, and biodiesel. Additionally, methanol is used as blend component to gasoline (mainly in China). Table 21 summarizes information on methanol fuel.

Table 21: Fact sheet methanol (based on DBFZ database 2016 as well as Majer (2010), Sakar (2011), Tunå (2014), Andersson (2014), Landälv (2016), and Hrbek (2016))

Methanol	Short description
Typical feedstock	Wood based black liquor, wood, solid organic waste (wood, municipal solid organic waste, straw), alternatively glycerol or biogas
Typical process	Mechanical pretreatment (crushing), drying, gasification, gas purification, gas treatment (reforming, water gas shift, sulphur removal, CO_2/water-removal), methanol synthesis, product purification (distillation); Alternatively by steam reforming of biogas for syngas production
Main technology provider (examples)	Regional focus: US, CN, EU Enerkem (methanol as byproduct, 1,000 t a^{-1}), BioMCN, Air Liquide, Haldor Topsøe, Mitsubishi Hitachi
TRL/FRL	4 to 6/5
Typical production costs	14 to 54 USD per GJ Mainly capital investment and feedstock price
Typical GHG	2 to 58 kg CO_{2eq} GJ^{-1} (according RED)
S | Strength	High knock resistance (high octane number) enable higher compression ratio and efficiency of spark ignition engines; free of aromatic compounds; Reduction of limited exhaust raw emissions because of high oxygen content
W | Weaknesses	High toxic chemical substance; material compatibility of fuel-carrying components sometimes not assured; low calorific value compared to conventional gasoline; low

		vapor pressure and high evaporation heat cause poor ignition characteristics at low temperatures; hygroscopic
O	Opportunities		Efficient synthesis, multipurpose fuel (e.g. as basic chemical, intermediate for different conversions to gasoline or diesel substitutes like DME or OME), direct application in fuel cells; alternative fuel component for diesel and gasoline engines
T	Threats		Toxic properties of methanol, blend walls and acceptance of methanol as fuel compared to ethanol as fuel, methanol derived fuels (DME, OME)

Zhen and Wang (2015) also describe different engine/fuel concepts for using methanol in combustion engines. Mainly discussed are the usage as pure methanol and in blends with gasoline in SI engines as well as in methanol/diesel fuel blends and as an addition to the intake air in CI engines. Binary blends of methanol with biodiesel, hydrogen, DME, and LPG, and ternary blends with other alcohols in diesel or gasoline fuels are also described in the literature (Zhen and Wang, 2015).

Emissions

Most studies have investigated the use of pure methanol or its blends in different engines and associated exhaust emissions. Unfortunately, non-standardized emission test procedures have generally been used. Therefore, only emission trends can be summarized from reported results.

Çay et al. (2013) compared emissions from SI engines fueled with methanol or gasoline, finding CO and HC emissions reductions of up to approximately 80% using methanol. Çelik et al. (2011) investigated use of pure methanol as fuel at high compression ratio in a single cylinder gasoline engine. In addition to increased engine power and brake thermal efficiency, they found a benefit in lower CO and NO_X emissions however HC emissions increased.

Several authors (Shenghue et al. 2007, Canakci et al. 2013, Han 2017) investigated 5% to 30% blends of methanol in gasoline in SI engines and passenger cars, testing only a small number of load points or constant speeds. A direct comparison of these results is not possible because of the different test setups. In general, it can be stated that blending methanol into gasoline has a positive effect on regulated emissions.

Methanol-diesel fuel blends have been investigated intensively. Sayin (2010) tested 5% and 10% blends in a single cylinder diesel engine at five engine speeds, finding decreased CO and HC emissions and lower smoke opacity. However, NO_X emissions increased by up to 50%.

Zhang et al. (2009, 2011) tested the addition of methanol to engine intake air. Using the Japanese 13 mode test and evaluating mixtures with 0%, 10%, 20%, and 30% methanol, they found increased CO and THC emissions in the raw exhaust gas. Adding a diesel oxidation catalyst (DOC) reduced the increase, so that no effect could be detected. NO_X and PM emissions decreased by approximately 10% and 22%, respectively. In contrast, Geng et al. (2014) found higher PM emissions at 50% load when adding methanol to the intake air. Performance at low load levels confirmed the results of Zhang et al. (2011).

Yusri et al. (2017) also reviewed emissions characteristics from different studies of 5% to 30% methanol blends in CI engines, finding no clear trends for all regulated exhaust compounds.

5.5.2 Lignocellulosic ethanol

Alternatively to so-called conventional ethanol based on sugar or starch crops, for which more than 70 million tons were produced globally in 2016, advanced ethanol is based on the use of lignocellulosic materials (biomass). Currently, the installed capacities for lignocellulosic ethanol production are about 0.5 million tons per annum, and several more production plants are in planning (representing another about ≈ 3.5 million tons per annum capacity). (Naumann 2016) Most lignocellulosic ethanol production is based on biochemical conversion via fermentation. However, thermo-chemical production of bioethanol from lignocelluloses via gasification and alcohol synthesis is also possible, as is syngas fermentation. Compared to the fermentation routes, these processes are not presently a large focus of international RD&D.

Most commonly ethanol is used as blend component in gasoline. But also pure ethanol can be used as fuel. In Brazil hydrous ethanol is used as pure fuel, which can mixed in flexi fuel vehicles with gasoline ethanol blends (Costa und Sodré 2010; Kyriakides et al. 2013; Melo, Tadeu C. Cordeiro de et al. 2012; Wang et al. 2015). Scania developed a diesel engine running on 95% ethanol and an ignition improver (ED95) (Stålhammar 2015; Velázquez et al. 2009). Yanowitz & McCormick (2016) published a review on practical issues of gasoline vehicle drivability. Table 22 summarizes information on ethanol biofuel.

Lignocellulosic ethanol is from the chemical side the same as first generation ethanol and should have the same emission behaviour. As there were no emission data found explicit to lignocellulosic ethanol production, emission data from other ethanol blends were used.

In comparison to neat gasoline, no clear trend can be observed for all emissions. This demonstrates that the emissions of ethanol blends are strongly dependent on the engine technology and operation. Figure 6 shows a summary of emissions test with different test cycles of ethanol blends.

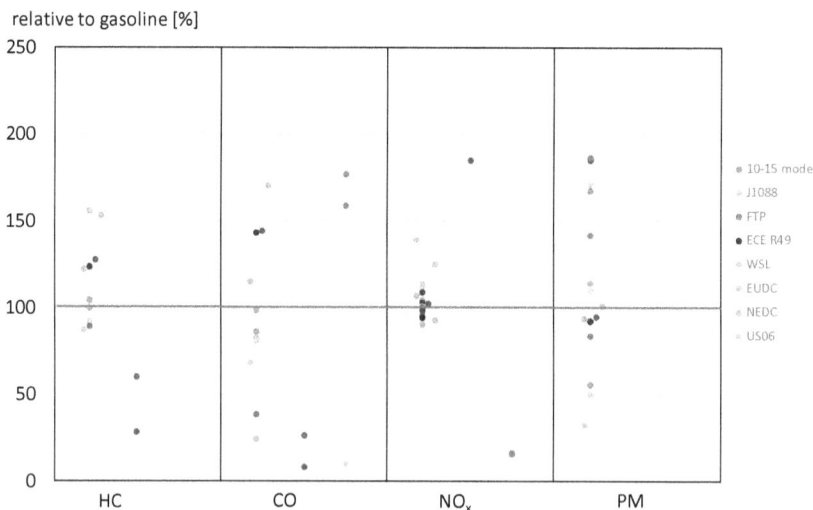

Figure 6: Overview for regulated emissions of ethanol blends relative to gasoline. For each emission component listed, the blend proportion starts with 5% on the left side and goes up to 85% on the right side. The single tests are coloured by the test cycles used.

Table 22: Fact sheet lignocellulosic ethanol (based on DBFZ database 2016 as well as Mueller-Langer (2014), Jeswani (2015), Gubicza (2016), Gerbrandt (2016), and Zech 2016)

Lignocellulosic Ethanol	Short description
Typical feedstock	Lignocelluloses (primarily straw, corn stover, bagasse, wood, switch grass); in case of syngas fermentation also organic (municipal) waste, industrial waste gas streams
Typical process	Biochemical pathway: pre-treatment (thermal, acid, etc.), hydrolysis, saccharification, C6/C5 fermentation, distillation, final dehydration with byproducts like lignin or lignin-based by-products, pentoses, stillage products such as fertilizer, biogas/biomethane, technical CO_2 Thermo-chemical pathway: pretreatment (crushing, drying), gasification, gas treatment (reforming, water gas shift, sulphur removal, CO_2/water-removal), ethanol synthesis, purification (dewatering with molecular sieve, distillation) or Hybrid respectively syngas fermentation: gasification and gas conditioning to syngas, fermentation, distillation with byproducts depending on process conditions: alcohols, organic acids technical CO_2
Main technology provider (examples)	Single plant capacities about 0.01 to 0.1 million t a^{-1}, regional focus US, China, EU, Brazil Fermentation: Clariant, Inbicon, Poet-DSM, Biochemtex, Raizen, GranBio, Abengoa, Borregaard Syngas fermentation: Lanzatech, Synata Bio (former Coskata, focus chemicals)
TRL/FRL	7 to 8/7 for fermentation
Typical production costs	21 to 46 USD per GJ for fermentation Mainly capital investment and feedstock price as well as revenues for byproducts, specific TCI about 2,030 to 3,160 USD per kW fuel
Typical GHG	4 to 32 kg CO_{2eq} GJ^{-1} (according RED) for fermentation
S | Strength	Use of lignocelluosic residues, application as multifunctional fuel or intermediate chemical (e.g., for ETBE, Alcohol-to-Jet, ethylene); high knock resistance (high octane number) enable higher compression ratio and efficiency of spark ignition engines; free of aromatic compounds; reduction of limited exhaust raw emissions because of high oxygen content; free of aromatic compounds
W | Weaknesses	Competitiveness with conventional ethanol in already large capacities worldwide; material compatibility of fuel-carrying components sometimes not assured; low calorific value compared to conventional gasoline; low vapor pressure and high evaporation heat cause poor ignition characteristics at low temperatures; hygroscopic
O | Opportunities	Improving value chain with regard to thin stillage use for biogas/biomethane and technical CO_2 (e.g. for industry or as renewable resource for synthetic power-to-fuels like discussed in the EU); also alternative blend/ fuel for diesel fuel (e.g., ED95 with 95% ethanol and 5% ignition improver)
T | Threats	Blend walls to E5/E10/E20 in most of the regions.

Rajesh Kumar and Saravanan (2016) review higher aliphatic alcohol applications in diesel engines, summarizing performance and emission results from neat and blended C3 to C12 aliphatic alcohols. Beneficial results are reported for particulate emissions. NO_x emissions generally decreased with propanol or butanol but increased with pentanol and higher alcohols in the blend. HC and CO emissions also increase with alcohols. The adoption of EGR and injection timing allows more emission improvements.

Giakoumis et al. (2013) review ethanol and n-butanol diesel blend transient emissions, with special emphasis on transient operations like cold/hot start, idle-acceleration-idle, smoke test, etc. The overall tendencies point to PM and CO reductions, HC increases with alcohol additions (possibly related to lower exhaust gas temperature reducing DOC efficiency), mixed outcomes for NO_x depending on specific conditions, and higher emissions of non-regulated BTX, PAHs and carbonyls.

Ghadikolaei (2016) comprehensively review blend and fumigation applications of methanol and ethanol fuels for CI, SI and motorcycle engines and their impact on regulated emissions, also providing extensive discussion on unregulated emissions (alcohols, alkenes, alkyl nitrites, peroxyacetylnitrate, polycyclic aromatic compounds (PAC), carbonylics, quinones, NO_2, N_2O, SO_2, metal species, dioxins) as well as giving individual summaries of selected referenced articles.

Karavalakis et al. (2015b) report on emissions from nine different gasoline vehicles running on ten different ethanol and isobutanol blends. Higher alcohol content leads to lower CO, CO_2, PM and PN emissions. On the other hand, THC, NMHC, CH_4, and NO_x did not exhibit any strong trends, and carbonylics generally showed higher emissions.

Schifter et al. (2013) performed a systematic study on gasoline blended with 10%, 20%, 30% and 40% v v^{-1} ethanol, both in anhydrous and 4% hydrous form, on a single-cylinder engine. Gasoline quality was a home-made blend from five commercial refinery streams (reformate, alkylate, catalytic cracker gasoline, isomerate and straight run gasoline). Particulates were not evaluated, but CO, HC and NO_x emissions throughout displayed slight decreases both with hydrous and anhydrous ethanol addition, indicating robustness towards small water quantities and a possible optimum regarding criteria emissions at a level of 20-30% ethanol.

Koivisto et al. (2015a) provide a systematic study on alcohol functionality in CI engines and resulting exhaust gas quality; see also two corresponding investigations on carbonyl/ether compounds and alkybenzene type fuels by Koivisto et al. (2015b, 2016).

More literature on neat and blend applications of aliphatic alcohols up to C_8 is compiled in Table 23.

Aliphatic alcohols: engine performance and emission issues (citation; title)

Zhao et al. (2010)	Carbonyl compound emissions from passenger cars fueled with methanol/gasoline blends
Zhao et al. (2011a)	Effects of different mixing ratios on emissions from passenger cars fueled with methanol/gasoline blends
Broustail et al. (2012)	Comparison of regulated and non-regulated pollutants with iso-octane/butanol and iso-octane/ethanol blends in a port-fuel injection Spark-Ignition engine
Daniel et al. (2012)	Dual-Injection as a Knock Mitigation Strategy Using Pure Ethanol and Methanol
Thomas et al. (2012)	Fuel Economy and Emissions of a Vehicle Equipped with an Aftermarket Flexible-Fuel Conversion Kit
Chen et al. (2013)	Effects of port fuel injection (PFI) of n-butanol and EGR on combustion and emissions of a direct injection diesel engine
Dai et al. (2013)	Investigation on characteristics of exhaust and evaporative emissions from passenger cars fueled with gasoline/methanol blends
Gu et al. (2013)	Experimental study on the performance of and emissions from a low-speed light-duty diesel engine fueled with n-butanol-diesel and isobutanol-diesel blends
Deep et al. (2014)	Assessment of the Performance and Emission Characteristics of 1-Octanol/Diesel Fuel Blends in a Water Cooled Compression Ignition Engine (SAE 2014-01-2830)
Britto et al. (2015)	Emission analysis of a Diesel Engine Operating in Diesel-Ethanol Dual-Fuel mode
Chan, T.W. (2015)	The Impact of Isobutanol and Ethanol on Gasoline Fuel Properties and Black Carbon Emissions from Two Light-Duty Gasoline Vehicles
Poitras et al. (2015)	Impact of Ethanol and Isobutanol Gasoline Blends on Emissions from a Closed-Loop Small Spark-Ignited Engine
Zheng, Li et al. (2015b)	Experimental study on diesel conventional and low temperature combustion by fueling four isomers of butanol
Zhang et al. (2016)	Effect of using butanol and octanol isomers on engine performance of steady state and cold start ability in different types of Diesel engines

5.6 Liquefied biomethane (Bio-LNG or LBM)

The use of liquefied natural gas (LNG) as transport fuel is increasing internationally. A renewable blending or substitution using liquefied biomethane (Bio-LNG or LBM) is part of RD&D especially in Europe where there is a growing interest in cross-border biomethane trade. Table 24 summarizes information on Bio-LNG.

Table 24: Fact sheet Bio-LNG (based on DBFZ data base 2016 as well as van der Gaag 2012, Steinigeweg 2015, Ricardo 2016, Scholwin 2017)

Bio-LNG	Short description
Typical feedstock	Food, agriculture and forest residues (straw, wood chips), solid municipal waste (landfill waste gas)
Typical process	Bio-chemical pathway: Pretreatment, anaerobic digestion, CO_2 removal (e.g. by pressure swing adsorption, membrane or cryogenic separation of CH_4 and CO_2) and liquefaction; (less relevant yet) thermo-chemical pathway via synthetic natural gas: pretreatment (crushing, drying), gasification, gas treatment (reforming, water gas shift, Sulphur removal, CO_2/water-removal), methane synthesis, purification, liquefaction
Main technology provider	Bio-LNG Wijster in the Netherlands (Rolande LNG and partners), Osomo, French CRYO-PUR®, Norwegian Campi AS/Wärtsila, Waste Management/Linde (US, landfill waste gas)
TRL/FRL	7 to 9/7
Typical production costs	13 to 16 USD per GJ for biowaste and landfill gas as biogas resource
Typical GHG	11 to 21 kg CO_{2eq} per GJ for biowaste and landfill gas as biogas resource (according RED)
S | Strength	Established technologies; reduction of limited emissions (NO_x or PM); reduction of carbon dioxide emissions; free of aromatic compounds; high knock resistance (high octane number) enable higher compression ratio and efficiency of spark ignition engines; Bio-LNG use without blending or to improve the quality of fossil LNG
W | Weaknesses	High energy needs for liquefaction; high methane slip; high material stress because of no use of additives and high combustion temperature
O | Opportunities	Transportable biomethane for areas without gas grid, upcoming marine LNG engine systems; alternative fuel for gasoline and diesel engines
T | Threats	Competitive use of biomethane in different sectors (depending on policies often CHP)

5.7   Biodiesel (FAME)

Fatty acid methyl ester (FAME) biodiesel is well investigated and is not a drop-in advanced biofuel. Nevertheless, this fuel remains relevant to substitute fossil diesel fuel. Therefore, a separate section on this fuel is included in this report for highlighting the newest developments and research results.

The reputation of 1st generation FAME biodiesel has worsened since its climax around the turn of the millennium because of the food versus fuel use conflicts and, regarding public reception of pros and cons, failure to discern disadvantageous engine emission data for neat vegetable oils (pure plant oils, PPO) from highly advantageous emission data for FAME biodiesel, which is a processed conversion product of virgin or used oils. One of the main reasons for adverse emissions from PPOs is their significantly higher viscosity as compared to most other road vehicle fuels. Spray formation and proper mixing of cylinder charge with intake air is hindered for viscous fuels, leading to less complete combustion and elevated soot emissions.

FAME biodiesel per definition and as specified by standards is a mixture of long-chain carbonic acid esters with simple lower alcohols, in EU regulation limited to methyl esters. Limitation to these simple esters for commercial use was agreed upon because suitability of other esters for engine use bears uncertainties and risks due to unknown effects of long-term engine use and lacking results from laboratory testing. Nevertheless, a broad range of esters bearing other acid and alcohol moieties have been tested and shown to be compatible with current engine technology, with broader application so far restricted by the higher production costs of these specialized chemicals.

Based on ever-greater experience with combustion of neat and blended biodiesel in vehicle engines, knowledge on respective engine adjustments and underlying combustion characteristics has grown continuously. The same holds true for controlling biodiesel quality by identifying and eliminating unfavorable constituents and impurities and by providing feedstock with adequately balanced fatty ester chain composition. The advent and growing use of waste oils and fats for biodiesel production pose new challenges on purification and elimination of incompatible components, which has been investigated and reported on by Kleinová et al. (2013).

Quality control of biodiesel and its blends with common diesel fuels is documented in several overviews available on-line, e.g.:

- Biodiesel Guidelines, WWFC 2009:

  http://www.acea.be/uploads/publications/20090423_B100_Guideline.pdf

- Biodiesel – Handling and Use Guide, 4th ed. Revised, NREL 2009:

  http://biodiesel.org/docs/using-hotline/nrel-handling-and-use.pdf

- Biodiesel.org, keyword search on biodiesel topics:

  http://biodiesel.org/what-is-biodiesel/reports-database

Hoekman and others, available as CRC Report No. AVFL-17a (2011), have published a comprehensive compilation of biodiesel properties and related issues.

Biodiesel properties are mainly determined by structural elements of the acid chain involved, with a relatively uniform range of chain lengths from C_{14} to C_{20} according to vegetable oil feedstock. First to mention among common concerns is storage stability and fuel blend integrity. Prominent factors influencing the oxidative stability of FAME biodiesel include its degree of unsaturation as well as the potential presence of bisallylic methylene carbon atoms (situated between two C=C-double bonds, "skipped dienes"). Radicalic and oxidative attack lead to spitting or coupling of carbon chains as well as potentially introducing new, possibly incompatible functionalities at the unsaturated sites. As a result, liberation of acidic degradation products, acceleration of reactions leading to compositional changes, and formation of undesired higher molecular products (oligomers) with limited solubility are observed. Viscous and insoluble components will disturb proper function of the fuel system and negatively affect fuel atomization in the engine cylinder, leading to suboptimal combustion.

On the other hand, unsaturated molecular structures provide lower fuel viscosity and sufficiently depress onset temperatures for parameters associated with cold flow. Here, pour point (PP), cloud point (CP), and cold filter plugging point (CFPP) are the biodiesel properties of greatest interest.

The boiling point of biodiesel, well above 300 °C, precludes combustion in spark ignition engines, which means that ignition behavior is described in terms of cetane rating. Generally, cetane numbers of typical constituents like straight-chain C_{16} to C_{18} esters are sufficiently high for proper combustion and may even serve as cetane improvers upon blending. Care has to be taken, however, in case of large proportions of unsaturated and/or branched chain components introduced by unusual feedstock. This topic has been investigated specifically for esters by Knothe et al. (2003) and Knothe (2014), and for many other substance classes including esters by Yanowitz et al. (2014).

Altogether there is a close interdependence of viscosity, cold flow properties and cetane rating which relate to the particular carbon chain structure of FAME biodiesel.

In order to extend the spectrum of possible biodiesel formulations, modification of the alcohol moiety of FAME has to be considered and has been performed; see the investigation by Cardoso et al. (2014). Even more ambitious is the concept of catalytically mediated "juggling" of ester components as described by Dubey and Khaskin (2016), who propose to apply ruthenium catalysts in order to exchange acid and alcohol moieties among esters (including triglycerides) while introducing new components. Such reaction sequences are of metathesis type in the sense of "substituent interchange" but do not alter sites of unsaturation, as will be introduced below as olefin metathesis.

Excursus: metathesis biodiesel

Among diesel fuel types, biodiesel displays a distinct boiling curve feature that makes a difference: it has a narrow boiling point range mainly covering temperatures near the upper margin defined in diesel fuel specifications. The lack of low- to medium-temperature boiling constituents is not as much a problem for combustion kinetics, but it is for engine oil dilution. Sweeping of cylinder walls by the piston constantly transfers small amounts of unburned fuel into the oil reservoir, from which it would evaporate during engine operation, provided the sump temperature is sufficiently high. Evaporation of mainly high-boiling biodiesel is retarded, leading to progressive oil dilution and risk of oil sludge formation by physicochemical degradation. This drawback can be mitigated or even eliminated by a reforming step called *olefin metathesis*, which comprises catalytic reactions developed decades ago.

In chemistry, the term "metathesis" generally describes a bimolecular interchange of groups at a distinct molecular site, which in case of an olefinic starting material is an exchange of substituents attached to a C=C-double bond. Conversion is accomplished by catalytic action of certain precious metals, most prominently ruthenium, which has been established after fundamental investigations by Grubbs, Schrock, Chauvin and others (see e.g. Grubbs (2004)) and has received major attention in olefin chemistry (see e.g. a review by Mol (2004)). Reaction sequences can be performed under mild conditions (slightly elevated temperatures, no pressure, no aggressive solvents) and may involve self-metathesis of starting material or cross-metathesis by applying olefinic co-reactants. The overall degree of unsaturation remains unaltered and cis-/trans-configuration of double bonds normally will be equalized, that is, the reaction is not specific towards a desired configuration. While this lack of stereospecificity is disadvantageous in synthesis of specific compounds, it is of minor importance for fuel reforming as long as final fuel properties do not suffer from altered cis-/trans ratios. If in doubt, one should consider literature data on melting points and cetane numbers of the respective isomers.

The starting material for fuel metathesis may either be neat vegetable oil or FAME biodiesel, though the latter normally will be preferred because it contains fewer impurities. The results of Nickel et al. (2012) indicate it is advantageous to remove hydroperoxides or other products of beginning alkenoic ester oxidation by applying a magnesium silicate purification step prior to catalyst application. These authors assume reactive oxidation products are able to inhibit catalyst activity. For recent perspectives on bio feedstock, refer to Jenkins et al. (2015), Mc Ilrath et al. (2015) and O'Neil et al. (2015).

Metathesis catalyst research does not only exploit commonly used ethylene as co-reactant, but also heavier olefinic hydrocarbons which introduce new structural elements among reaction products. Recent investigations of importance have been published by Montenegro and Meier (2012); Vorfalt (2010; Ph.D. in German on synthesis and mechanisms of ruthenium-NHC-complexes for olefin metathesis); Pillai et al. (2013) and Dobereiner et al. (2014). The latter work, describing a tandem isomerization/metathesis strategy, as well as work from Nickel et al. 2012 cited above, shed light on how catalysis research explores possible pathways for substance alterations, which in case of biodiesel substrates might establish options to simultaneously alter chain length distributions and modify positions and/or configurations of C=C double bonds. This would be an interesting approach to reducing or eliminating sensitive bis-allylic structural features, which act as triggers for thermo-oxidative degradation, without reducing the otherwise valuable degree of unsaturation. Preparative chemists can follow ongoing research topics at

http://allthingsmetathesis.com/.

Since only unsaturated compounds will react, highly saturated oils/biodiesels are not of much interest for this process. The palette of substances obtained will grow substantially, comprising both olefinic species and unsaturated esters. Because naturally occurring acids/esters from plant oils display inner double bond positions, a high proportion of the conversion products will have reduced chain lengths compared to the starting material. The resulting boiling curve of metathesis fuels thus is lowered, and it is smoothened due to the extended range of substances, which can be regarded as a convenient and effective mixture of oxygenated and oxygen-free organics.

Whether or not the product is suitable for proper engine operation has to be elucidated by laboratory analyses of relevant parameters and by performing engine tests. Literature on metathesis fuel characterization is scarce and even more investigations on engine emission testing are published. To be highlighted are a 2014 Ph.D. from Pabst (2014, in German, on emission measurements with diesel fuel, biodiesel, HVO/HEFA and metathesis fuels), Munack et al. (2012) and Schröder et al. (2016).

5.8   Fuel properties: tabulated analysis data

Fuel parameters covered by published fuel specifications represent minimum requirements, in some cases subject to regional legislative adaptions. Many more additional parameters have been established together with corresponding analytical methods. Customers, fuel providers and dealers, engineers and scientific staff may need such additional fuel data for special purposes. We made an attempt to collect information on fuel properties from published research articles (Table 25). Regulated parameters such as water content, ash, and oxidation stability are not specific for a fuel. These quality parameters differ from batch to batch and depend on handling and storage of the fuels. Therefore, they are not included in Table 25.

Table 25: Selected neat fuel properties from research literature. The values result from the references listed above in this chapter.

	HVO/HEFA	FT fuel	Biodiesel	OME (OME3/OME4)	DME	Methanol	Ethanol
Cetane number [-]	80-110	>74	47-66	78/90	55-60	5	8
Octane number Research [-]						106	107
Octane number Engine [-]						92	89
Density (15 °C) [kg m^{-3}]	≈780	≈780	883	1,030/1,070	667	791	789
Flash point [°C]	80-94	100-120	>150	54/88	-42	11	17
Lubricity (HFRR) [μm]	≈420		320	534/465		1100	1057
Viscosity at 40 °C [mm^2 s^{-1}]	2.9-3.0		≈4.5	1.05/1.75 (25 °C)	<0.1	0.58	1.13
Carbon [% m m^{-1}]	≈85	≈85		44.1/43.4	52	37.5	52
Hydrogen [% m m^{-1}]	≈15	≈15		8.8/8.4	13	12.5	13
Oxygen [% m m^{-1}]	0		10.3	47.1/48.2	35	50	35
specif. heat of evaporation [kJ kg^{-1}]					467	1163	918
lower heating value [MJ kg^{-1}]				19.1/18.4	28.8	19.6	26.8
Boiling Point [°C]			≈360	155/200	-25	64.7	78.3

6 Chemical reactions among fuels' components and additives

6.1 Introduction to fuel stability and chemical fuel reactions

Among organic combustible chemicals, engine fuels show comparatively moderate to low reactivity at normal ambient conditions. Fuel composition as obtained from the supplier will only change to a minor degree at storage in tanks or vehicle fuel systems. Nevertheless, since fuel components come with structural moieties that are susceptible to chemical modification and degradative attack, fuel quality deterioration and reduced shelf life may occur. Accelerating factors are elevated temperature, irradiation, oxygen or oxidizing agents, radicalic species and catalytically active trace metals. The general propensity of fuels to oxidative degradation is reflected by the fact that all common market fuels have to fulfill oxidation stability requirements from respective standards as noted in Chapter 4.

Changes in chemical nature of fuels become evident in several ways, such as changes in color, formation of opaque, amorphous or crystalline deposits, and measurable modifications of physical or chemical parameters that not necessarily are perceivable macroscopically. Resulting negative impacts from such processes may be poor operation or malfunction of pumps, filters, injectors, valves, pistons and exhaust equipment, but also impairing of engine combustion, emission levels and engine oil quality.

Fuel types most often associated with deterioration and deposit formation are biodiesel and diesel-biodiesel blends, provoked by unsaturated esters as specific structural elements. Ordinary hydrocarbons bearing such olefinic structures would experience similar conversion reactions, but in general saturated, aromatic and naphthenic hydrocarbons are subject to degradation too.

Laboratory parameters which serve to indicate degradation propensity, undesirable impurities or reaction products, are C/H/O-ratio, olefinics and aromatics content, iodine number, oxidation or storage stability, acid number, peroxide value, total contamination or water+sediment, carbon residue. All of these parameters refer to bulk properties, which means that they reflect an integrated sample property irrespective of what distinct chemical species contributes to this factor. To undertake such compound-specific assignments, application of special analytical methods is necessary (IR, NMR, fluorescence spectroscopy, MS, sensor techniques), mostly in conjunction with chromatographic separation to reduce sample complexity.

Stringent quality control at fuel production, distribution and storage is indispensable to provide optimum shelf life and maintain fuel specification for extended time periods. But, since fuel reactivity cannot be brought to zero, fuel quality will fade with time and requires special attention in engine applications with inherently longer storage periods, such as PHEV, stand-by power generators or recreational boats. Under such circumstances, microbial attack is another important mode of action responsible for fuel degradation and especially formation of sludge and deposits. Elevated water content of fuels and poor housekeeping at fuel distribution and storage are important factors promoting microbial growth. Microorganisms especially reside and grow at fuel-water interfaces, build up biofilms, produce and excrete reactive metabolic products which promotes further degradation and corrosion. The resulting biofilm microcosm is a viscous sludge that plugs filters, injectors etc. and finally causes instrumental failure. A broad spectrum of aerobic and anaerobic microorganisms is involved.

The complexity of microbiological characterization and substrate metabolism cannot be elucidated here. Readers interested in this topic might start with remarks given by Jakeria et al. (2014) in a review on factors affecting biodiesel stability. To get an impression on microbial fuel deterioration and applied methods, see specific investigations described by Raikos et al. (2011) on aviation fuels; Zimmer et al. (2013) on diesel-biodiesel B0, B7, B10, B100; and Bücker et al. (2014) on diesel-biodiesel blend B10 and the effect of biocides. Some common antioxidants usually applied to suppress oxidative degradation could prove to suppress microbial growth too, see an investigation by Beker et al. (2016).

Most important among fuel degradation issues are stability and turnover of biodiesel and its blends with common, nonpolar fuels. Corresponding investigations and research results will therefore be a prominent part in this chapter.

6.2   Mechanisms and measurement of deposit formation in fuels and in the fuel system

6.2.1   Thermo-oxidative fuel degradation: oxidation and stability testing

Different fuel types, fuel applications and legislative demands have led to a range of test methods suitable for the determination of fuel behavior at thermal or thermo-oxidative stressing. Oxidative or storage stability lab testing mostly are accelerated methods in order to gain conclusive results in acceptable time. But for comparative purposes and to reflect real-world situations it is also useful to simply apply distinct ambient conditions over prolonged storage periods, as did Farahani et al. 2011 in a ten-month campaign using ULSD, three biodiesels, B5-B20 blends and ambient conditions ranging from -25 °C to +35 °C and 25 to 100% relative humidity, allowing intermittent access of air. Sample acid numbers remained nearly unaffected and water as well as sediment precipitations to a great extent were reversible, indicating response to annual temperature/humidity change. 1H-NMR indicated saturated carbon chains preferably appearing in precipitates, but negligible fuel degradation. Yang, Hollebone et al. (2014) subjected canola and soy biodiesel and its ULSD blends to one-year storage experiments at 15 and 40 °C excluding air and light but adding copper, water and copper+water. Copper promoted degradation, while water did not; significance of acid value, induction time and ester chain constituents slightly depended on impurities added. Jose & Anand (2016) investigated highly unsaturated Karanja and highly saturated Coconut biodiesel in neat, mixed and fossil diesel-blended form over ten-month storage periods, which are each exposed to four different conditions concerning light/dark and air/no air. Unsaturation and access of light and air expectedly accelerated deterioration, which was somewhat slowed in Karanja-Coconut mixtures as compared to pure biodiesel type and biodiesel-fossil diesel blends. It can be speculated whether natural antioxidants in these biodiesels unintentionally provided a limited stability improvement.

Oxidation stability parameter of biodiesel and biodiesel blends as determined in fuel standards refers to results of accelerated aging procedures in which ambient air bubbling (Rancimat method) or pressurized oxygen (PetroOxy method) plus elevated temperatures are applied to fuel samples. Experimental data simply tells a time span for onset of degradation product formation (conductivity due to volatile acids) or an oxygen consumption threshold, respectively. A different method applicable to petroleum fuels is called Rapid Small Scale Oxidation Test (RSSOT), which has been applied to compare ULSD, biodiesel and blend stability (Dodos et al. 2014). Asadauskas & Erhan (2001) describe a thin-film test method at 150 °C to investigate oxypolymerization of vegetable oils and lubricants. Some methods on storage stability employ aeriation with or without irradiation and by providing fuel contact to an activated (polished) copper strip or spiral. Results obtained by such tests are reported as amount of solid products formed and weighed after isolation. Thermogravimetric analysis (TGA) is used to monitor thermal behavior of fuels. Results to some extent are related to oxidative stability (see an article by Jain & Sharma 2012), but from much higher temperatures employed, they are especially important with regard to thermal stressing in common-rail fuel recirculation, engine oil deterioration and coking of engine parts. Lu & Chai (2011) and a Ph.D. by Chai (2012) describe biodiesel pyrolysis at temperatures from 200 to 900 °C and application of TGA and a semi-isothermal tubular flow reactor for product and mechanism investigations.

A report by Westbrook (2005) notes on comparison of biodiesel oxidation stability test methods, Knothe (2006a) draws attention to install reproducible conditions for fuel sample contact surface with oxygen and reported measurable data deviations. Jain & Sharma (2010) published a general overview on relevant methods for biodiesel characterization and laboratory methods. Pullen & Saeed (2012) wrote a thorough review on biodiesel oxidation stability addressing adverse effects and chemistry of biodiesel oxidation, parameters, measurement methods and influencing factors. Yaakob et al. (2014) provide a review on biodiesel stability parameters, testing methods, issues of biodiesel standards and discuss effects of antioxidants and metals. Botella et al. (2014) investigated a suite of twelve widely differing biodiesels to compare Rancimat and PetroOxy stability testing and arrived at correlating data from both methods. 4-allyl-2,6-dimethoxyphenol and catechol were tested as stability additives and performed well in all cases. Kivevele et al. (2015b) provide an outline on biodiesel stability issues and introduce a suite of less common non-edible vegetable oil biodiesels of African origin in this context.

Christensen & McCormick (2014) made use of storage stability method ASTM D 4625 (43 °C, restricted oxygen admittance) to simulate clean 21 °C underground storage over months to years. Fresh and pre-aged B100 and biodiesel blends were subjected to this procedure and revealed biodiesel with high bisallylic character to be most problematic for long-term storage, but overall good stability results with respect to fuel standards. Addition of antioxidants to pre-aged biodiesel partly re-established/maintained stability, showing that long-term storage of FAME and its blends is possible under clean and properly controlled conditions. Lab testing control should better rely on oxidation stability methods than determination of acid and peroxide value, since these latter will indicate property changes only *after* onset of degradation reactions.

Østerstrøm et al. (2016) performed extended aging experiments on a commercially produced B30 RME-diesel blend containing 2-tert-butylhydroquinone as synthetic antioxidant. Time course covered 43 days at 90 °C and 58 days at 70 °C, each applying air flows of 50, 100, and 150 ml min^{-1}; conditions were intended to subject samples to high or worst case stressing. Typical parameters like peroxide value (PV), TAN and induction period (IP, remaining stability buffer) were monitored together with sample mass changes and FAME ester profiles. Air flow rates had minor effects, while samples held at 90 °C showed markedly earlier onset and steepness of induction period declining and corresponding TAN and peroxide data. Depletion of unsaturated esters started after 20 days at 70 °C and after about five days at 90 °C (with only minor losses of saturated chains), paralleling PV and TAN and the diminishing stability buffer as indicated by IP data.

6.2.2   Molecular structures involved in thermo-oxidative biodiesel degradation

Generally, fuel deterioration test methods mentioned in Section 6.2.1 measure global parameters, but do not intend to clarify mechanistic details of attack or identify molecular species involved in degradation processes. To prevent fuel deterioration by taking appropriate stabilization measures and to understand underlying mechanisms, elucidation of molecular species was performed and still is an important research topic; for early work see e.g. a report to CRC and NREL by Waynick (2005). Investigations go parallel both in fuel research and in corresponding fields of dietary science targeting vegetable oils and lipids. For examples of application of instrumental methods for monitoring involved chemical species in oxidative conversion see Guillén & Ruiz (2004, 2005) on 1H-NMR targeting aldehydes in oxidized unsaturated oils and Hayati et al. (2005) on FT-IR targeting peroxides in oxidized oil-water emulsions.

Fang & McCormick (2006) published a fundamental investigation on degradation pathways in biodiesel-diesel blends. By applying FT-IR, NMR and gravimetric analysis they drew conclusions on how radically initiated attack on different parts of biodiesel ester chains may proceed to degradation products, oligomers and deposits. Bisallylic -CH_2- positions of polyunsaturated esters are especially prone to radicalic H-abstraction, which is directly related to oxidation stability and reflected in FAME fuel specification parameter "polyunsaturated FAME $\geq$4 C=C double bonds". A second important mechanism involves aldol condensation among aldehydes or ketones (oxidation products themselves) and esters leading to α,β-unsaturated esters. Water traces and iron species promote degradation. Deposit formation is most severe at around B20 due to most insufficient solvation power of nonpolar diesel towards polar products in this constellation, with low-sulfur diesels displaying the lowest solvation power.

The outstanding importance of allylic and bisallylic carbon chain structures for the stability of FAME of various origin was described by Iyer (2016) in a recent review referring to methyl oleate/methyl linoleate mixtures and some FAME types.

Chuck et al. (2012) investigated aging of neat biodiesel by applying UV-VIS spectroscopy, FT-IR spectroscopy, GC/MS, 1-H and 13-C NMR as instrumental methods to detect reactive intermediates and to trace changes of certain structural elements (olefinic, bisallylic, aldehydes, branched acids) over a 360 h time period. Aging temperatures were 90 °C and 150 °C, leading to higher viscosity and higher molecular weight oligomers (determined via size-exclusion chromatography) at 150 °C. Oligomer formation is assumed to proceed mainly via epoxide rather than peroxide intermediates, a C18 ketone was determined as subsequent degradation product. Formation of aldehydes and C1-C3 acids were monitored. As several other authors have noted too, course of reaction pathways and products depends on temperature applied (e.g. Singer & Rühe 2014). Morales et al. (2014) performed aging of vegetable oils at 40 °C while monitoring diminishing tocopherol (natural antioxidant) levels and reported the occurrence of hydroperoxydienes; these have been assumed as intermediates but most probably will be missed when higher aging temperatures are applied. Altogether, these investigations emphasize the difficulties in predicting and explaining fuel degradation processes encountering different thermal conditions; correspondingly, the degree of fuel deterioration occurring in an individual vehicle will reflect the specific thermal history of its tank-filling batch.

Flitsch et al. (2014) from biodiesel Rancimat tests provide quantitative results for aging products that have been proposed or detected by previous work. Determination of formic and acetic acid was accomplished by sampling both fuel reaction batch and trap water sample over a twelve hours period with hourly intervals, showing formic acid as dominant volatile acid. About half of acetic acid formed remained in the fuel reaction batch. The time course of C5 through C18 fatty acid formation was monitored over a 50 h period and correlated to reaction times exceeding oxidation induction periods (IP). Lower free fatty acids may be formed from corresponding aldehydes, which themselves originate from oxidative chain cleavage via hydroperoxides; higher fatty acids to some extent stem from acidic hydrolysis of starting methyl esters. Two epoxy compounds, cis- and trans-epoxy stearic acid methyl ester, have been identified as oxidation product detectable after IP in amounts as much as some few mass-%.

Kleinová et al. (2013) demonstrated that using waste cooking oil (WCO) as biodiesel feedstock will introduce analogous degradation and oligomer type species into finished (esterified) WCOME as would occur on oxidation of fresh biodiesel produced from non-used oils, unless WCOME is subjected to adequate purification workup. NIR/MIR spectra are used to characterize sample oligomer status and assignment is supported by spectra of methyl ester of a distilled synthetic dimeric fatty acid.

Thermal reaction sequences without essential interference of oxygen go parallel to thermo-oxidative processes, which means that the degree of oxygen supply will influence the product spectrum obtained. Destaillats & Angers in two publications (2005a, 2005b) report on formation of cyclic and bicyclic fatty acid monomers and conjugated isomers of linoleic acid in heated edible oils. Both *intra*molecular cyclization and *inter*molecular chain coupling reactions may occur. While such carbon-carbon couplings are possible via radicalic processes, additional pathways are opened after thermally induced C=C-double bond isomerization from bisallylic (skipped) dienes to conjugated dienes. These latter undergo Diels-Alder coupling with an olefinic carbon double bond forming six-membered carbon ring structures, see notes in the review by Jakeria et al. (2014) and a Ph.D. by Vukeya (2015) describing different temperature regimes for fuel stressing (from typical storage to pyrolysis conditions) with detailed discussion of possible degradation products.

Photo-oxidative fuel degradation partly encounters pathways different from thermal and thermo-oxidative pathways, but leads to corresponding oxygenated compounds like alcohols, aldehydes, ketones and acids. Sample irradiation therefore has to be controlled and taken into account when interpreting the product spectrum obtained from aging experiments or analyses of fuel batches with unknown history. Choe & Min (2006) discuss implications of various stressing conditions, together with mechanisms of stabilizer action, in a review on edible oils, the findings and statements being likewise applicable to fuel esters since acid moiety structural features are concerned.

6.2.3 Other issues of fuel degradation

In order to better resolve structural features of reaction intermediates and degradation products, it is convenient to employ model compounds representing typical structural elements for experimental investigations, as did Lu & Chai (2011), Chai (2012) and Iyer (2016) for main C18-biodiesel esters (methyl stearate, -oleate, -linoleate). Vukeya (2015) performed thermal and thermo-oxidative stressing experiments using n-hexadecane, tetralin and decalin as oxygen-free model compounds. See also investigations by Naziri et al. (2014) on squalene and squalane.

It has been discussed whether reaction intermediates or products of fuel composition bear catalytic potential with respect to accelerating fuel degradation. If this were the case, deteriorated fuel remnants in a vehicle tank would spoil freshly added fuel at refilling, and thus continuously worsen fuel quality with time similar to carryover of substrate inoculation by microorganisms. According to results of Strömberg et al. (2013), this is not the case; authors conclude "… primary and secondary products do not directly influence the degradation rates." Instead, refueling would dilute degraded components and thus improve quality of aged fuels. Possible secondary effects, though, are taken into account such as enhanced metal mobilization due to acid generation and continuous buildup of water traces from increasing hygroscopicity of aged biodiesels. Results were obtained from accelerated aging of RME B100 and reference fuel B7 at 80 °C with monitoring IR spectra and water content, and determination of lower acids by ion chromatography.

Saturated and unsaturated aldehydes are products of oxidative degradation of hydrocarbon and biodiesel fuels following decay of peroxide/hydroperoxide intermediates. Lu & Chai (2011), Chai (2012) and Vukeya (2015) have performed determination and mechanistic considerations for engine fuel model compounds, but aldehyde species have long been detected both in stored or oxidized biodiesels and in edible oils. Of special concern are unsaturated aldehydes due to their irritating and suspected genotoxic properties, see publications by Guillén & Ruiz (2004) on hydroperoxy- and hydroxyalkenals from sesame oil and do N. Batista et al. (2015) on formation of hexanal, 2-heptenal and 2,4-decadienal from biodiesel. Some aldehydes like hexanal, 4-hydroxy-nonenal, and 2,4-decadienal are recognized to be frequent or regular oxidation products which serve to monitor oil and fat quality, complementing other chemical methods like peroxide value, anisidine or thiobarbituric acid reagents (see an investigation by Liu et al. 2014). α,β-unsaturated 4-hydroxy-aldehydes, toxic and bioreactive degradation intermediates of special concern, have been determined by Han & Csallany (2009) in heated samples (185° C; 0, 1, 2, 3, 4, 5, 6 h) of single C18:0 through C18:3 methyl esters representing most common FAME. C18:0 and C18:1 FAME did not generate these aldehydes, while from C18:2 and C18:3 the hydroxylated hexenal, octenal and nonenal species were detected in considerable amounts. 2,4-decadienal appeared as another degradation product, but seemed not to be a precursor to hydroxy-alkenals.

Oxygenate ethers are not subject to degradation processes discussed above for hydrocarbon species, but display some reactions to be considered. Formation of explosive ether peroxides in long-term storage with sufficient access of oxygen is a commonly known hazard for some aliphatic ethers like diethylether (DEE). Peroxide formation can be suppressed by antioxidants like sterically hindered phenols, which find their application in hydrocarbon fuel stabilization too. DME as emerging engine fuel is considered much less prone to peroxide formation (see literature notes given in Section 5.3) than OME. OME, though, bear the structural element of an acetal and thus are subject to hydrolysis when exposed to hydrous acids (see e.g. in a Ph.D. by Lautenschütz 2015). Such conditions may occur in fuel formulations with elevated trace water amounts and unsaturated fuel components that experienced degradation stress generating organic acids. Hydrolysis propensity is an inherent feature of OME type fuels corresponding to their synthesis via catalytic equilibrium reaction of formaldehyde(-precursors) and methanol, and it is the reason for inclusion of water, methanol and methylformate parameters in OME quality control (applying to DME specification too). OME hydrolysis in aqueous acidic media was investigated by Baranton et al. (2013) for OME2-3 in 0,1m $HClO_4$ (perchloric acid) between room temperature and 90 °C showing rapid turnover. Suggested extrapolation from OME1-3 results point to increasing susceptibility to acidic hydrolysis for higher OME homologues. Zheng, Tang et al. (2015a) inferred water induced hydrolysis of DMM and OME from methanol and formaldehyde appearance in reaction mixtures. Their investigations, designed to improve synthesis of higher OME from DMM (syn. OME1) and paraformaldehyde by means of acidic ion-exchange resins, included determination of rate constants for propagation (-CH_2O- insertion for next higher homologue) and depolymerization at 60 °C, 70 °C and 80 °C and kinetic modeling. Schmitz et al. (2015) described corresponding issues of chemical equilibrium among OME synthesis, employing aqueous solutions of formaldehyde-methanol as starting mixtures.

6.2.4   Effects of various fuel components on deposit formation

Processing of vegetable oils and other feedstock to obtain biodiesel must include removal of distinct impurities, of which monoglycerides, especially those carrying a saturated fatty acid chain (saturated monoglycerides, SMG) are of concern. Monoglycerides have been regulated in biodiesel standards to obey maximum mass-% values of 0.4 (ASTM D 6751) or 0.7 (EN 14214), together with maximum contents for di- and triglycerides, free and total glycerin. Phytosterols bound to carbohydrates, called steryl glycosides, SG (or more specifically glucosides, if bound to glucose) are similarly regarded as unfavorable impurities, but do not appear as biodiesel specs parameter. Both substance groups mentioned can cause precipitation above the cloud point – corresponding to regular crystallization – of fuels, an effect that to some extent presumably corresponds to higher polarity exerted by free hydroxyl groups of glycerin and carbohydrate structural units in SMG and SG. Camerlynck et al. (2012) investigated the effects of both these trace impurities from a practical perspective of filterability and critical content levels by performing cold soak experiments and scanning electron microscopy. Chupka et al. (2011) provide an overview of reported crystallization and filter blocking observations attributed to SG and SMG impurities and performed a thorough investigation on the effects of cooling rate and temperature history on biodiesel(-blend)/SMG precipitation. It was concluded that slower real-world cooling rates (as compared to lab testing) and extended periods of diurnal temperature cycles generate larger and more stable crystals with higher filter blocking potential and retarded crystal melting even upon heating much above cloud point temperatures. Crystal phase polymorphism of SMG and individual eutectic points of SMG with different FAME types have to be considered. McCormick (2012) has published a short summary of this work as *NREL Highlight*. Fersner & Galante-Fox (2014) report on bench-top filtration testing of different biodiesels doped with SG, SMG and carboxylate salts. Filterability results depended on biodiesel feedstock and are regarded as controllable by deposit control additives.

Oxidation inhibitors, antioxidants, stabilizers – Biodiesel purification inevitably eliminates most or all of the biogenic oxidation inhibitors like tocopherols that are natural secondary components of plant oils. Whether such natural antioxidants can be preserved to market-ready biodiesel in sufficient quantities and provide the desired activity for fuel applications would require narrow-spaced controls while leaving risks for missing stability targets. Therefore, it is common practice to spike biodiesels (and fossil fuels, too) with specific amounts of synthetic stabilizers which act as radical scavengers (primary stabilizers), like 2,6-di-tert-butyl-4-hydroxytoluene (BHT). Secondary stabilizers like tris-nonylphenyl phosphite act as peroxide decomposers; application of both stabilizer types may prove beneficial in terms of fuel stability. Setup of stabilizer recipes requires fuel-specific adaptions to be aware of overdosing risks: A remark in the introduction of WWFC biodiesel guidelines

http://www.acea.be/uploads/publications/20090423_B100_Guideline.pdf

states "The overuse of anti-oxidants can lead to the additional formation of sludge"; a similar hint is given on p. 8 of the article from Botella et al. (2014). EN fuel standards (EN 14214, EN 590, EN 16734; see resources compiled in Chapter 4), while generally recommending stabilization by antioxidants, place corresponding warnings concerning possible deposit formation among arctic fuel classes due to antioxidant use. Possible reasons for such deposits are not specified further; warnings in fuel standards presumably reflect reported problems with suggested assignment to antioxidants. Though Schober & Mittelbach (2004) in an investigation on performance of synthetic phenolic antioxidants did not find negative impacts on CFPP and other parameters of different biodiesel types, it should be kept in mind that fuel providers use different concepts among antioxidant additives without allowing insight into individual speciation of additive packages.

The review by Choe & Min (2006) discusses different modes of oil degradation and terms of stabilizer action, the findings of which can be well ported to FAME as structurally related organics. Dwivedi & Sharma (2014) provide a review dealing with antioxidant and metal influence on biodiesel stability, Kivevele & Huan (2015) cover the same topic for biodiesels from two Eastern African non-edible oil feedstocks. From this and other literature, it appears that choice and dosage level of antioxidants may cause unexpected results including pro-oxidative effects. Different antioxidant types, e.g. natural and synthetic, can annihilate their potentials instead of displaying synergism. The limited value of natural antioxidants in fuel storage applications must be looked at in the context of their original function, that is to eliminate adverse effects on storage lipids at ambient temperatures and humidity, intermittent irradiation, and in conjunction with the biogenic speciation of respective plant materials.

Cetane improvers (CN boosters) like alkylnitrates, azo compounds or di-tert-butyl peroxide are sometimes used as fuel additive in order to enhance ignitibility (cetane rating) of diesel fuels. The underlying principle is to generate radicalic species at elevated temperatures to initiate radical chains, an effect which will proceed to a small extent even at ambient temperatures. From this, it can be derived that cetane boosters inevitably contribute to fuel deterioration during storage and thus counteract the protective function of antioxidant/stabilizers added. Such evidence is discussed e.g. by Ribeiro et al. (2007) and Fang et al. (2003); the latter authors even propose a test procedure for fuel oxidation stability making use of alkyl nitrate addition.

Formation of deposits and sludge in fuels usually is regarded as ordinary physical precipitation due to insufficient solvancy power of the fluid. Some substances, however, pose a higher potential to form viscous or thickened masses by so-called self-structuring properties; they act as organogelators (organic mass gelators). This means small amounts as low as 1% are able to bind or intercalate molecules of the surrounding fluid, even simple solvents like tetrachloromethane, in larger structural units. Though these usually are not stable at higher temperatures and will release bound substances reversibly, some quite stable and highly viscous formulations exist. Research in this field to a great extent occurs in dietary, oil, pharmaceutical and cosmetics sciences, where preparation of viscous products is a definite target; while in fuel science we deal with unwanted effects, effects follow the same principles in fuels. It is thus feasible to have a look over the "scientific garden fence". For example, the extensive article by Terech & Weiss (1997) on "Low Molecular Mass Gelators of Organic Liquids and the Properties of Their Gels" is cited quite often in this field.

A single substance most prominent as organogelator is 12-hydroxy-stearic acid (HSA) and its lithium salt. Carboxylate salts are known for their soap forming properties, but HSA especially is a thickening agent used commercially in lubricating greases. It is produced in large amounts from the corresponding mono-unsaturated 12-hydroxy-octadecenoic acid from castor oil. It can be assumed that analogous processes in FAME type fuels may produce hydroxylated products that possess some gelatin functionalities contributing to higher viscosity in oxidized fuels.

Finally, we want to recall the simple mechanism of regular physical crystallization of fuel constituents as represented by cold-flow properties. This does also include additional filter blocking effects by ice crystal formation in water-contaminated fuels. Tucker et al. (2016) give a comprehensive report on such low temperature operability issues.

7 Known health effects: emission testing, data and relevance

7.1 Health effects of engine emissions

The adverse effects of engine emissions are common knowledge and have been documented in numerous studies, leaving still many unresolved questions concerning modes of pollutant inventory buildup and impacts on susceptible organisms. Additionally, particulate and gaseous exhaust components undergo complex alterations after liberation into ambient air, generating a wide spectrum of potentially hazardous material. This has been noted by IEA-AMF as a future task to deal with in their strategic plan 2015-2019 (5th Strategic Plan for the Implementing Agreement (IA) for Advanced Motor Fuels (AMF):

"With respect to concerns related to pollutant emissions, research has been increasingly focusing on secondary aerosols. Secondary aerosols have a significant impact on human health, climate, and air quality. They are formed from fine atmospheric particulate matter (composed of a complex mixture of organic and inorganic materials) and oxidation of volatile precursors. AMF considers the evaluation of the impact of secondary aerosols on human lungs to be an important topic for the future. [...]"

Although of general importance, dealing with turnover of air pollutants is an issue of atmospheric and environmental chemistry, which is far beyond the scope of this study. Here we focus on investigations dealing with measurement and evaluation of motor emissions captured by way of engine-out raw exhaust, tailpipe, dilution tunnel, upstream or downstream of aftertreatment devices, eventually referring to investigations on in-situ (in-cylinder) spectroscopic methods. Data on regulated and non-regulated emission profiles thus obtained provide information on primary pollutant sources associated with vehicle traffic, serving at understanding one important aspect of air pollution research. It is at this point to recognize the clamp to ambient aerosols: fuel type, engine type, emission tier level, aftertreatment functionality and test cycle/driving conditions are of major importance for the amount and composition of gaseous and particulate primary emissions, which subsequently affects chemistry and fate of secondary aerosols.

To get an impression on the subject of secondary aerosols in relation to engine exhaust have a look at some of the following exemplary research results. Gupta & Singh (2016) describe compositional analysis of engine combustion particulates and their ambient fate. Bahreini et al. (2012) and Gentner et al. (2012) worked on discrimination of gasoline and diesel emissions and their relative contributions to ambient secondary aerosol pollution. Brines et al. (2015) proved that ultrafine particle pollution in large urbanized cities experiencing intense solar irradiation can be assigned mainly to nucleation of traffic exhaust components, using cluster analysis and inspection of diurnal pollution profiles. An extended general study on interrelating vehicle emissions and ambient air quality was undertaken for CRC and has been condensed in a final report by Robinson (2014, additional database material available online). Platt et al. (2013), Gordon et al. (2014) and Liu et al. (2015) performed reaction chamber experiments for detailed inspections of transformation processes of fresh engine exhaust contributing to secondary aerosol, of which Gordon et al. specifically focused on the impact of exhaust aftertreatment, fuel chemistry and driving cycle on resulting aerosol character. Aimanant & Ziemann (2013) studied transformation of diluted n-pentadecane vapor as model compound by subjecting it to OH radicals and NO$_X$ in a reaction chamber and detailed characterization of transformation species. Agarwal et al. (2013) evaluated comparative toxicity of biodiesel and common diesel nanoparticles with benefits for biodiesel, as determined from surface area distribution, elemental/organic carbon and PAH content.

Research on sources and fate of ambient aerosol needs sophisticated sampling techniques and microscopic and physicochemical characterization of particulate fractions and species. Not surprisingly, great experimental overlap exists between exhaust gas sampling and ambient aerosol research. Literature citations highlighting these topics are included in Section 7.1.2 and with further reading in 7.1.3. The reader will learn about specifics on nucleation mode, Aitken nuclei, accumulation mode and "coarse" fraction (≈ 1 μm) particles as relevant size classes in gaseous-liquid and gaseous-solid equilibria and turnover processes.

Among non-regulated parameters, the most difficult to assess and at the same time of highest concern are mutagenic and carcinogenic effects of particulates. Even more challenging and thus scarce are monitoring results for cytotoxic effects attributable to exhaust gas. Upon inhalation, NO_x and carbonylics act as irritants and via prolonged exposure may cause carcinogenic or necrotic effects as well. Narcotic, nephrotoxic or other adverse effects of engine exhaust are difficult to detect and therefore not much covered by studies. Over the course of the descriptive overview following in Sections 7.1.1 and 7.1.2, reference to investigations and reviews on health effects associated with vehicle emissions are made, complemented by more citations and hints on information resources given in the further reading in Section 7.1.3.

Health risks arising from skin contact and inhalation of unburnt fuel are mainly questions of public and occupational health and thus will not be dealt with here. Nevertheless, as stated in Section 4.3 (Fuel regulation and fuel standards) such issues of course have to be kept in mind in order to prevent people, biota and ecosystems from health hazards and pollution.

7.1.1 Harmful gaseous constituents of engine exhaust

Though engine combustion emissions comprise a wealth of individual organic and inorganic compounds, to date regulatory measures (criteria emissions) still apply to just two single species and two bulk parameters each: Carbon monoxide (CO) and nitrogen oxide NO_x specifically; total hydrocarbons (HC, occasionally THC) and particle amount (PM, PN) as bulk parameters. Inspection of particle number distributions and extension to particle sizes as low as a few nanometers is a very recent development in regulatory determinations.

Historically, <u>CO</u> has been a more severe problem associated with engine exhaust than it is today, as combustion efficiency together with usage of oxygenated fuel additives remarkably augmented the situation. The effect of CO to induce inner suffocation by blocking oxygen transport capability of blood is common knowledge and follows a well-understood mechanism. Toxic effects of nitrogen oxides <u>NO_x</u>, in contrast, cover several points of attack in living organisms, the mechanisms, severity and implications of which still remain controversial and a matter of research. Most often, NO_x emission data (and data on ambient NO_x levels too) implicitly refer to the sum of nitric oxide NO and nitrogen dioxide NO_2, the fractions of which can be obtained separately from raw data of the chemiluminescence measurement. To lump together $NO + NO_2$ from combustion emission testing is legitimate because NO is rapidly converted to NO_2 in ambient air and itself proved to pose far less toxic hazards than NO_2. Known acute effects are irritation of mucous epithelia and disturbance of lung function at concentration levels around one to few ppm with markedly lower effect levels for persons with asthmatic disposition. Indications for genotoxicity exist, but no clear evidence for carcinogenic or other chronic effects has been attained. Epidemiological studies attempting to elaborate clear indications in this respect always suffer from various exposure factors contributing to possible manifestations. To get some information on nitrogen oxide toxicology see e.g. a review by Latza et al. (2009) on experimental and epidemiological studies and EU recommendations on occupational exposure limits (SCOEL/SUM/53 – Nitrogen Dioxide, The Scientific Committee on Occupational Exposure Limits 2014), (SCOEL/SUM/89 – Nitrogen Monoxide, The Scientific Committee on Occupational Exposure Limits 2014).

One important issue of adverse NO_x effects is its reactivity towards PAH and particulate carbonaceous material from combustion leading to nitrated species R-NO_2 (see e.g. Adelhelm et al. 2008), which proved to bear higher carcinogenic potential than corresponding base structures. Jariyasopit et al. (2014) and Bocchi et al., (2016) recently published research results demonstrating such evidence.

Gaseous engine emissions of total hydrocarbons (THC) are legislatively capped and commonly measured by unspecific flame ionization detection (FID). The test result is considered a bulk parameter and will include contents of most types of volatile, nonpolar, combustible C-bearing species, irrespective of their environmental or toxicological implications. In fact, species that are detected this way belong to substance classes denoted in environmental sciences as volatile organic compounds (VOC). In cases where concentrations of methane (CH_4) are to be determined separately, the remainder hydrocarbons are referred to as non-methane organic gases (NMOG) or non-methane [volatile] hydrocarbons (NM[V]HC). To discern reactive or toxic species of special interest, specific sampling and separation techniques like GC/MS are necessary. The respective laborious workup is one reason for individual hydrocarbons still being not regulated by emission standards. From a health and environmental protection perspective such implementation is desirable, since VOC proved to be reactive hydrocarbons in atmospheric processes producing hazardous haze and smog. Recent work by Roy et al. (2016) and Drozd et al. (2016) gives an impression on relative amounts of different volatile organics in gasoline exhaust gas depending on vehicle technology, driving conditions and ambient temperature. The results emphasize the outstanding impact of cold start engine emissions both in terms of relative substance class contributions and absolute amounts. Accordingly, high percentages of cold start vehicle operation among overall fleet mileage will annihilate standard estimates of emitted pollutants commonly derived from averaged test cycles and type approvals. This adds to underestimates due to emission testing discrepancies between dynamometer or test bench runs and real driving cycles.

Most hydrocarbons are susceptible to photochemical formation of radicalic intermediates that subsequently react with atmospheric oxygen, which in turn forms ozone via processes involving hydroxy and hydroperoxy radicals. The term *reactive hydrocarbons* thus points to their ozone forming potential and, regarding the consequences of the photooxidative attack involved, at the same time to their role as precursors or "seeds" for oligomeric material, which is a prerequisite to the buildup of secondary aerosols. Here, beside VOC also heavier semi-volatile organics (SVOC) and primary PM from combustion enter the scene and contribute to growth and transformation processes of aerosol particulate matter.

In addition to reactive volatile hydrocarbons, formaldehyde and NO_x have been identified as prominent promoters of ozone formation. Accordingly, Annex X of the European Air Quality Directive (Directive 2008/50/EC, European Parliament and the Council 2008) lists about 30 compounds recommended for monitoring, comprising C2 to C8 aliphatic/unsaturated hydrocarbons, BTX aromatics plus formaldehyde plus NO_x. Singular substances among the ozone precursor list at the same time are considered hazardous in a general sense, exerting more or less severe direct effects. Most prominently, benzene as known carcinogen and 1,3-butadiene are to be mentioned in this respect. The latter has narcotic and irritating properties and is classified as carcinogenic (Institute for Health and Consumer Protection, European Chemicals Bureau 2002); see also EU SCOEL recommendations (SCOEL/SUM/75 – 1,3-butadiene, The Scientific Committee on Occupational Exposure Limits 2007). Exhaust gases from gasoline engines operated with high-olefinic fuels display elevated butadiene levels (Hajbabaei et al. 2013) which underscores the definition of a maximum allowable olefin content by gasoline specifications.

Volatile carbonylic substances deserve special attention in terms of toxicology and atmospheric aerosols, with aldehydes being of more concern than ketones. Analytical determination can be performed by GC/MS together with ozone precursor screening, but more frequently they are trapped by specific solid phase derivatization cartridges using 2,4-dinitrophenyl¬hydrazine (or equivalently, by bubbling exhaust gas samples through a reagent solution). The corresponding hydrazones thus formed are eluted from the cartridge and quantified by means of RP-HPLC. Target compounds include alkanoic, alkenoic and aromatic aldehydes and ketones. Care has to be taken to avoid artifacts with acetone data, since adsorption cartridges are easily contaminated by acetone present in laboratory fumes from solvent use. Aromatic ketones or quinones usually are detected only in minor amounts in vehicle exhaust, but are frequently found in ambient aerosols due to photooxidative processes.

Engine exhaust research performed over the years does not provide a clear picture on carbonyl emissions, but formaldehyde, the simplest aldehyde $H_2C=O$, consistently appears as prominent or even major single compound among this substance class. It displays strong irritating, skin-sensitizing and muta-/carcinogenic potency and contributes to atmospheric ozone and haze. It is an important intermediate species of fuel combustion and is routinely monitored in kinetic studies of cylinder charge ignition and combustion (e.g. Mayo & Boehman 2015, Skeen et al. 2015). Elevated formaldehyde emission levels will occur for methanol fuel combustion, unless engine adjustment and exhaust aftertreatment are state-of-the-art (see e.g. Zhang et al. (2011).

Other volatile aldehydes more or less abundant in exhaust gas are aliphatic homologues ranging from acetaldehyde to hexanaldehyde, acrolein, methacrolein, 2-butenal and simple aromatics like benzaldehyde or tolualdehyde, accompanied by minor amounts of simple aliphatic ketones. They represent partially oxidized derivatives of common fuel constituents or oxidized fragments thereof. An article by Fontaras et al. (2010) serves as example for emission measurements of carbonylics, showing no consistent trend for exhaust gas contents from biodiesel combustion. Authors also provided a compilation of aldehyde health impacts documented in literature. An evaluation of published results on biodiesel carbonyl emissions is included in CRC Report No. AVFL-17a by Hoekman et al. (2011); no clear trends could be stated which authors attribute to restricted data base, variability of testing conditions, possible deviations in aldehyde analysis and doubtful biodiesel quality. While generally combustion of fuels with high oxygenate share including biodiesel are suspected to emit slightly higher carbonyl emissions, distinct carbonyl species corresponding to the original fuel oxygenate, like acetaldehyde for ethanol fuels, can be expected to occur. Exhaust carbonylics content varies strongly with driving mode, and to rely on oxidation catalysts in final exhaust treatment could turn out to be fatal, as turnover efficiencies may be insufficient especially for acetaldehyde; see publications by Elghawi & Mayouf (2014) and Hasan et al. (2016). We hereby again recognize that emission testing should span more than just four criteria parameters to tell the whole story of assets and drawbacks of different fuels.

Among carbonyl species, those coming with the structural element of α,β-unsaturation (conjugated carbonylics, R-CH=CH-CH=O) are of special toxicological concern. Common features of these substances are a pungent smell and a possible muta-/carcinogenic potential due to their high reactivity towards DNA, either directly or after transformation to corresponding epoxides. Low-molecular weight monounsaturated species like acrolein, methacrolein or 2-butenal would be captured by standard workup via hydrazone derivatives, but larger molecules like trans-trans- (tt-) 2,4-decadienal (C10) have been detected which require a more laborious quantification. From food chemistry research, many of these higher mono- and polyunsaturated carbonylics are recognized both as natural flavoring principles and undesirable products of oil- and fat deterioration, for which a causal relation can be assumed, i.e. oxidative modification of fatty ester chains.

Bearing this in mind, it comes as no surprise that conjugated carbonylics not only occur in stored biodiesels and have been detected in laboratory pyrolysis products of oils and FAME (see Chapter 6 on fuel reactions), but have also been identified in biodiesel exhaust (Yang et al. 2007, Wu & Lin 2012). Whether they originate from deteriorated biodiesel by escaping engine combustion as "survivors" or occur because of incomplete combustion, awaits clarification. In any case, it draws attention to the necessity to control toxic aldehyde contents of stored biodiesels and especially those made from used cooking oils (UCOME) prior to their fuel utilization.

7.1.2 Harmful particulate constituents of engine exhaust

Numerous severe health effects of engine exhaust correspond to particulate matter emissions, and we will have a closer look at toxicological implications, measurement and characterization of finely dispersed aerosol matter.

Particle sizes liberated by combustion engines have decreased substantially over the years due to changes in engine design and requirements of engine power and efficiency. Modes of emitted particle sizes today overwhelmingly cover the sub-micron range; we thus are talking about nanoparticles. These are of special concern as they endanger functionality of cells and organisms simply due to this minority, allowing them to pass biological barriers like epithelial cells or mucous membranes and thus to invade tissues and body fluids like blood or lymph. An investigation by Hudda & Fruin (2016) addresses this topic by applying a diffusion charging instrument to assign particle types associated with alveolar lung deposition risks, as part of size distribution monitoring by means of CPC and SMPS instruments in order to apportion ambient aerosol inventories to relevant emission sources. A commentary on the need to pay special attention to smallest sizes, i.e. sub-10 nm, was placed in Particle and Fibre Toxicology by Pedata et al. (2015).

Next to deal with, particle surfaces surely are not inert but display a heterogeneous patchwork of more or less active sites carrying poisonous or reactive substances, multiplying their negative impact. Given the large variability in particle surface chemistry depending on fuel type and modus of combustion – how to predict and explain interactions and noxious effects on biota? Investigations *in vivo* are not the only way to gain results, since artificial test systems mimicking biotic functionalities have been developed. For instance, Penconek & Moskal (2016) make use of artificial mucus in order to elucidate factors affecting the interaction of diesel exhaust particles (DEP) with lung epithelial mucus layers. Bisig et al. (2016) make use of a special cell culture termed multi-cellular human lung model which proved to assign a range of toxic effects to diesel exhaust gas, while no such effects were detected if emissions from a flex-fuel engine operated with E0, E10 and E85 fuel were applied to the test system. In a 2009 paper Vouitsis et al., based on eco-toxicity testing results from bacteria bioluminescence bioassay, already concluded that focus on simple exhaust PM reductions will not adequately reflect toxicity potentials. This is due to declining overall particle mass paralleling a shift to smaller, but more hazardous particles, and specific toxic particle burdens from polycyclic, metals, etc.

The potential and ambivalence of particle filters, catalysts and other strategies of exhaust aftertreatment with respect to health hazards is a much more complex topic than suggested, because a simple "before-after" budget of target species may overlook secondary effects or non-criteria components. Some reported examples will be cited in further reading Section 7.1.3; generally it is problematic to compare results obtained from different sampling points along the exhaust line or from vehicles employing different exhaust treatment techniques. Electron microscopic characterization is a powerful tool to monitor alteration of particle properties as early as primary aerosol still dwells the vehicle exhaust system, as for example Liati et al. (2013) demonstrated in terms of particulate composition, structure and reactivity when passing filters and catalysts. In a later publication, Liati et al. (2016) describe distinct electron microscopic differences among primary 4-55 nm particles from GDI vehicle emissions. Degree of crystallinity, reactivity and surface area parameter vary with engine status and are suspected to correlate with health hazards.

A historically established parameter relating to the amount of fine particles is smoke opacity or *(Bosch)filter smoke number (FSN)*, simply denoted as "smoke" or "soot" in older literature (see online *DieselNet.com* resource

https://www.dieselnet.com/tech/measure_opacity.php

for a technical overview). It uses either light attenuation when passing exhaust gas through a sample volume of defined length (*continuous opacity*) or reflectivity of a filter paper before and after charging with exhaust particulates (*spot opacity*); see a methodical investigation by Lapuerta et al. (2005). Determination of particle mass (PN) is performed by passing diluted exhaust gas through Teflon coated glass fiber filters kept at 51.7 °C (125 °F), which is to control bias resulting from condensed semi-volatile exhaust components. Investigations have been performed to establish correlations between opacity, FSN and particle mass (Northrop et al., 2011, testing diesel-biodiesel blends and looking at carbon fractions), but for general mechanisms of light scattering with very small and irregular particles, the opacity principle has serious drawbacks for accurate determination of high proportions of nanoparticulate material. However, since significant improvements have been made to establish modern smokemeters, optical smoke opacity is a reliable test method still used in mobile control campaigns such as "Snap Acceleration Test" (current legal application example, New Hampshire, URL:

http://www.des.nh.gov/organization/divisions/air/tsb/tps/msp/documents/diesel_opacity.pdf)

and moreover is performed for comparative purposes in lab emission testing (see e.g. Armas et al. 2014).

Measurement of particle number (PN) and size distributions requires impactors and particle sizers able to rapidly scan the size spectrum of interest like EEPS, FMPS, SMPS, ELPI; see work on instrumentation comparability (mobility particle sizers, diffusion chargers) by Kaminski et al. (2013) and a comparison of performance and application to exhaust, ambient and artificial aerosol by Zimmermann et al. (2014). More literature on PN determination are provided in Sections 7.1.3 and in 7.2.

Following measurement of particle mass, particle number and particle number concentrations (size distributions), a first characterization of exhaust particulates would describe major chemical fractions.

Beside soluble and insoluble organics, subfractions to be discerned are soot, carbonaceous material, elemental and organic carbon; concerning inorganic species, ash content, trace metals, sulfate, nitrate or other oxidized compounds are recognized, see a description by Gupta & Singh (2016). Trace metals are of special relevance with respect to toxicological implications and functionality of engine parts and aftertreatment devices. In exhaust soot they occur as oxidic or sulfatic ash originating from inherent fuel content, or from intentional additivation which targets engine oil functionality, enhanced fuel combustion and catalytic diesel particulate filter regeneration (see more remarks on trace metals and metalloid additives in Section 7.2.3)

As noted in the preceding Section 7.1.1, part of the genotoxic potential of engine exhaust comes with gaseous components like carbonylics and volatile hydrocarbons, but most important with respect to mutagenic or carcinogenic effects are exhaust particulates. These health risks are associated with soluble organic matter extractable with appropriate solvents, recovered as *soluble organic fraction* (SOF), an established parameter generally applied over the course of mutagenicity/carcinogenicity testing. To focus on effects and chemical constituents of SOF correlates with the fact that exposed tissues of organisms will only suffer from liberated reactive components, that is, if molecular species of concern are transferred from particle surfaces to surrounding aqueous fluids and biological transport factors. Organic solvents usually applied in the laboratory are assumed more effective for extraction of nonpolar organics than tissue fluids, hence fractions obtained by laboratory workup should always reflect an upper limit of toxic potential of a particulate sample. To attain a complete picture, it is recommended to sample both particulate and semi-volatile fractions of engine exhaust, the latter being recovered as liquid condensate by means of a condenser placed downstream of the particulate filter. For both matrices, dichloromethane (DCM) is considered as the most suitable solvent for extraction of nonpolar genotoxic principles.

Causalities of carcinogenicity are difficult to verify, since living species are subject to a variety of adverse effects that could cause cell degeneracy. Knowledge largely stems from epidemiology and animal exposure campaigns. Testing for genotoxic and mutagenic effects, in contrast, requires far less time for significant results, which can be attained by standardized laboratory procedures using cell cultures like *Ames* Test with TA98 strain of *Salmonella typhimurium*. For details on strategies and methods in exhaust toxicology please refer to dedicated literature given below in Section 7.1.3 – further reading.

Chemical species first and foremost responsible for genotoxic effects associated with engine exhaust particles are polycyclic aromatic hydrocarbons (PAH), their isologues containing O-, N- and S-heteroatoms, polycyclic species with nitro- or amino substituents and respective homologues with partial structures displaying saturated sidechains or rings. Altogether, these can be denoted as polycyclic aromatic compounds (PAC), with PAH providing the majority of exhaust polycyclics. Among lower PAH, acute toxicity becomes a more prominent feature than muta-/carcinogenicity, while the reverse is observed for higher homologues. The condensed carbon rings of PAH represent cutouts of the graphitic C-network and are considered soot precursors with structural analogies explaining their co-occurrence and efficient adsorption on exhaust soot particle surfaces. The interrelationship between carbonaceous species, soot and PAH has been studied in lab burner experiments and correlated to engine combustion; see publications by Maricq (2011; 2014) as examples.

Aromatics present in fuels provoke higher levels of exhaust PM, black carbon and PAH (Karavalakis et al. 2015a, Yinhui et al. 2016). Relative influence of gasoline aromatics is more accentuated than for diesels due to generally higher soot and PAH production of the diesel process. Polyaromatics may pose genotoxic effects directly, but very often, they do so after metabolic activation, which means formation of epoxides and/or hydroxyl derivatives (the same applies to simple aromatics as small as benzene). These metabolites are intermediates of detoxification processes intended by living organisms to convert nonpolar substances into water soluble, thus extractable species – a strategy that unintentionally worsens the problem. Similar oxygenated polycyclic are formed via biotic and abiotic oxidation in air, water and soil.

Manzetti (2012) in a *letter to the editor* points to another suspected mode of action of PAH: Their structural resemblance to steroidal hormones, which in case of interference with biochemical regulation of humans (or, organisms in general) would pose additional threats to health.

7.1.3 Exhaust gas toxicology and characterization of particulates

The European Commission's Scientific Committee on Occupational Exposure Limits (SCOEL)

> http://ec.europa.eu/social/main.jsp?catId=148&intPageId=684&langId=en

holds a list of hazardous substance monographs in the form of recommendations on occupational exposure limits via URL

> http://ec.europa.eu/social/BlobServlet?docId=3803&langId=en.

Listed entries of special interest for fuel and combustion matters are DME, acrolein, nitrogen dioxide/nitrogen monoxide, methyl formate, 1,3-butadiene, naphthalene, MTBE, formaldehyde, benzene, mineral oil aerosols and 2-butenal.

McClellan et al. (2012) give a comprehensive review on past and current legislation on vehicle emissions and ambient air quality, studies on associated health effects and differences in diesel exhaust from traditional and recent technology, which should be taken into account by making sound distinctions in research and regulation. They examine detailed results for chemical species determined in exhaust gas, referred to in this paper. Manzetti & Andersen (2015) provide a review on emission products from bioethanol and ethanol-gasoline blends, the implications of additives and minor common oxygenates on health, and regulatory consequences.

A succession of four annual surveys 2011-2014 on research and ongoing activities concerning engine exhaust toxicity has been prepared for AMF Annex 42 by Czerwinski (2011-2014). Each report provides explanatory chapters and visualizations, short summaries of activities performed in surveyed countries and abstracts or cover frames of relevant literature.

Special attention has to be drawn to occupational health in environments suffering from restricted ventilation and exchange of breathing air; see a presentation by Bugarski (2012) held at 14th U.S./North American Mine Ventilation Symposium, and a Ph.D. on diesel emissions, PAH and SVOC in South African mines performed by Geldenhuys (2014).

Madden (2016) undertook an analysis of toxicology reports on biodiesel combustion emissions published since 2007. Though contradictory to some extent, results indicate that exhaust from biodiesel has similar toxicological risks as from fossil diesel. Gamble et al. (2012) published a review on occupational epidemiology literature with critical remarks on methodology and conclusions drawn, stating that appraisals are not sufficient to clearly confirm the diesel-lung cancer hypothesis. To proceed in this respect, that is, to help in elucidation of diesel exhaust as significant or controlling impact factor for negative aftermaths possibly manifested later, correlation with biomarkers would be extremely valuable. The obstacles in establishing such indicators are addressed in a 2014 review by Morgott (2014) including remarks on diesel technology and tables listing the palette of PAH, oxy- and nitro-PAH, aldehydes and ketones identified in exhaust and aerosol ambient matter. Recent reviews on health risks associated with vehicle exhaust emissions and underlying mechanisms provide insight into current toxicological knowledge and research, e.g. Donaldson et al. (2009), Gilmour et al. (2015), Manzetti & Andersen (2016) and Steiner et al. (2016).

Some investigations on exhaust gases, their chemical and toxicological evaluation are to be mentioned relating to oxidative (redox) potential of particulates, which is an important parameter to infer specific threats by oxidative stress on biota, their cellular response and cytotoxicity. The mechanisms and conceptual lines of current understanding have been described by Andreau et al. (2012). Determination of respective properties in diesel exhaust have been undertaken by Guarieiro et al. (2014, PAH content, redox reactivity and carcinogenity calculation of neat and blended biodiesel nanoparticulates), Joeng et al. (2015, monitoring of NO_x, organic and elemental carbon, and cytotoxicity of diesel exhaust) and Godoi et al. (2016, measurement of oxidative potential of exhaust from combustion of ULSD and B20-ULSD by ESR). Godoi and collaborators found higher values of oxidative potential for B20 exhaust which correlated with amorphous carbon disorder and trace amounts of copper in particulates, showing best correlation among a list of trace elements measured. SCR aftertreatment proved to reduce oxidative potential of exhaust. Tang et al. (2012) reported adverse effects of short-term (two hours) diesel particulates in vitro exposure on human lung and lung carcinoma epithelial cells by applying confocal Raman spectroscopy, atomic force microscope (AFM) and cytokine/chemokine monitoring with multiplex ELISA.

Investigations attempting to unveil exhaust health effect trends from comparing fuels or engine machinery quite often come with unexpected results. For example, genotoxicity of PM extracts obtained from vehicle combustion of ULSD containing 10% n-butanol or hydrous ethanol were found to exhibit higher genotoxicity than from diesel fuel combustion itself (Cadrazco et al. (2016). If measurement strategy covers an extended set of biological endpoints, results may indicate benefits in one respect but potential hazards in another. See for example results reported by Gerlofs-Nijland et al. (2013) who employed pro-inflammatory mediators, cytotoxicity and oxidative potential as hazard indicators for diesel particulates and found "...that B50 fuel usage could be equal or more harmful compared to diesel despite the lower PM mass emission." (lower mass emission due to use of B50 instead of B0 and/or using DPF equipment). Bisig et al. (2015) investigated gasoline exhaust and found increased anti-oxidative response and proinflammatory potential in lung cell tests while aryl hydrocarbon receptor (redox) activation and genotoxicity was reduced if exhaust had passed a non-coated gasoline particle filter (GPF). Considering exhaust treatment technology, filter gas flow design, operation as catalytic or non-catalytic unit and catalyst speciation will make a difference in toxicological signature.

Understanding of aerosol pollutant inventories, their sources, fate, properties and modes of action requires further investigations beyond figures of mass and number. Detailed characterization of diesel exhaust particles is described by Alam et al. (2016) and Popovicheva et al. (2014) (2015), a comparison of compositional and structural differences between gasoline and diesel engine soot is published by Uy et al (2014) with special focus on particle robustness and hardness and its impact on engine oil and wear. Maricq (2014) investigated basics of combustion particulate speciation (black carbon and soot) from ethylene and propane lab flames and light duty vehicle exhaust considering issues of soot standardization and vehicle operation. Jin et al. (2016) examined a wide range of fresh primary emissions and ambient aerosol with respect to a general particle characterization and cytotoxicity.

Hygroscopicity of exhaust particles, measured via size growth tendency, depends on fuel type and combustion performance and is one of several factors that will control fate of ambient aerosol particulate matter. Recent investigations on this property in exhaust particulates have been performed by Happonen et al. (2013) and Popovicheva et al. (2015), revealing more hydrophilic or hygroscopic nature for particles from oxygen containing fuels (HVO/dipentylether or biodiesel in these cases) and correlated to water-soluble ashes or C-O structures in soot. It can be expected that hygroscopicity will also affect interaction of particles with cell surfaces, fluids and transport vectors in biota.

To elucidate compositional features of ambient and combustion particulates, specialized spectrometer techniques were developed like the *Single-Particle Mass Spectrometry (SPMS)* technique (not to be confused with *Scanning Mobility Particle Sizer [SMPS]* instrument), as used by Vogt et al. (2003) and Lee et al. (2006). SPMS technique received improvements and today is considered as *Aerosol Time-of-flight Mass Spectrometer (ATOFMS)*, see application to diesel vehicle emissions by Toner et al (2006) and Shields et al. (2007).

With respect to continued reduction of exhaust particle sizes over the years, adaptions of detection optima of such spectrometers have taken place. A more recent development is *Single Particle Inductively Coupled Plasma-Mass Spectrometry (spICPMS)*, which is especially suitable for measuring heavy elements associated with soot, while the carbonaceous matrix is surpassed due to the intrinsic low sensitivity of ICP towards carbon. A dedicated instrument for aerosol soot measurements is called *Soot Particle Aerosol Mass Spectrometer (SP-AMS)*, see an introduction by Onasch et al. (2012). Dallmann et al. (2014) provide an example of SP-AMS application to engine exhaust characterization. One result is that engine oil contributes significantly to both gasoline and diesel vehicle exhaust which impedes engine type source apportionment in ambient pollution research.

A technical paper by Chow et al. (2008) gives a broad overview on applicable instrumentation for integrating, continuous, sizing and single particle measurement in air pollution research. More specific to engine exhaust emission testing, two SAE papers by Price et al. (2006) and Cavina et al. (2013) provide comparative investigations by applying different particle measurement instruments to gasoline DI engine exhaust.

7.2 Basis for health effect information: generation of emission data

7.2.1 General remarks on exhaust emission legislation and measurement

In order to elucidate the suitability of a fuel for broad future application in combustion engines, emission testing is of prime importance. Though much effort is spent to develop straightforward equipment and reproducible procedures, results gained from such investigations show remarkable scatter. This is a result of a large number of factors affecting combustion reaction kinetics and engine performance. The only way to gain reliable results on emission trends in fuel intercomparisons, or to gain data needed in vehicle approval procedures, is to apply standardized test procedures and to strictly follow prescribed procedural conditions. For different engine or vehicle types a range of applicable test protocols has been developed. To represent different driving situations, test protocols have to cover a minimum of low to high engine torque and speed settings, the combination and duration of which is tailored to intended engine/vehicle applications. The outcome of a data agglomeration of test cycle results for measurements comparing alternative and reference fuel emissions is part of Chapter 5.

It is important to note that valuable information on combustion behavior and engine performance cannot solely be reached by applying full, dedicated test cycles, but also by running a limited suite of representative load points or even constant settings. Investigations of this less comprehensive type, whether they are run on serial engines, prototypic or single-cylinder test engines, generate pilot data and are helpful to unveil mechanisms of combustion behavior. While these approaches are not intended to provide reliable data on real world engine emissions, this is not a serious drawback since even established test cycles are not able to do so. Discrepancies between emission data from standardized test bench protocols and real driving meanwhile attained common perception and have become a matter of public debate. This is triggered by scandals from exertion of defeat devices or defeat strategies – technical measures that are related to engine control units (ECU). Future investigations following real-driving emissions (RDE) test protocols will hopefully give a better picture and, together with durability regulations, promote the development of real ultra-low emission vehicles. To have a look on subjects of emission control, emission factors and RDE/PEMS development, refer e.g. to Weiss et al. (2012), Kousoulidou et al. (2013) and Mamakos et al. (2013a).

Instead of transferring details of test cycle and exhaust emission legislation from official documents to this report, we would like to introduce the complexity of this topic by suggesting to the reader a presentation by Hill (2013) covering more than 200 pages, entitled "Emissions Legislation Review", its subtitle includes the ironic plea "… and a sincere request not to shoot the messenger !", pointing to both the continuously launched amendments and the wealth of procedural details to cope with in emission testing. Requirements and procedural details are documented in U.S. Code of Federal Regulations, accessible via eCFR of "Title 40 — Protection of Environment / Chapter I —Environmental Protection Agency (continued) / Subchapter U — Air Pollution Controls", in both part 1065 (engine testing procedures),

http://www.ecfr.gov/cgi-bin/text-idx?SID=c1cdbcc08581e2c0675bab7e03bc2399&mc=true&tpl=/ecfrbrowse/Title40/40cfr1065_main_02.tpl

and part 1,066 (vehicle testing procedures),

http://www.ecfr.gov/cgi-bin/text-idx?SID=c1cdbcc08581e2c0675bab7e03bc2399&mc=true&tpl=/ecfrbrowse/Title40/40cfr1066_main_02.tpl

and corresponding parts referenced to therein.

Directives in force generally obey UN-ECE Regulation no. 83 for passenger cars and light duty vehicles and UN-ECE Regulation no. 49 for HDV. The respective amendments to a base directive have to be considered in emission testing if a specific engine type is to be aligned with regulatory requirements. Nevertheless, for R&D purposes any appropriate test procedure may be applied to whatever engine in order to determine and compare the influence of fuel type and driving points on exhaust quality.

To follow recent discussions on emission legislation, vehicle type approvals and corresponding technical requirements, have a look at: Pavlovic et al. (2016) concerning CO_2 emissions and energy demands mandatory for type approval; Zacharof et al. (2016) on type approval and real-world CO_2 and NO_x emissions; Ntziachristos et al. (2016) on diesel emissions control failures and consequences on emission factors and pollution control; and Mamakos et al. (2013b) dealing with accurate measurement of finest particles under regulated and non-regulated driving conditions (DPF regeneration or ambient -7 °C).

A booklet on heavy-duty and off-road vehicle test cycles, emission standards and reference fuels is furnished by Delphi Corp. (2016). The Dieselnet.com online portal lists information on emission test cycles used for emission certification and/or type approval, as well as seldom used or earlier test versions, as online resources accessible via

https://www.dieselnet.com/standards/cycles/.

Both resources provide visualizations of driving modes to be tested, like vehicle speed versus time (transient cycles) or engine speed/torque settings (stationary cycles). Published literature on emission testing investigations applying test cycles often comes with such plots to allow for better orientation. In cases where only a limited set of load points are occupied in stationary emission testing, it is very helpful if a corresponding graph is provided. This could be a complete test map valid for an approved cycle in which load points actually covered in the investigation are highlighted. Basically there is no restriction in applicability of test sequences to engine type: with stationary cycles, speed/torque settings of operation points are given as percentage of indicated engine specifications; with transient cycles, required vehicle speed and acceleration values are mostly moderate and thus piloting is easily achieved by modern vehicles.

7.2.2   Factors affecting measurement of engine emissions

In order to make correct correlations between changes in exhaust gas quality and variations in fuel composition it is important to be aware of the many factors and variables that affect exhaust measurement and generated parameter data. Factors and variables fall into two categories: those related to engine and fuel type, injection strategy and installed exhaust gas treatment, and those related to test cycle design and performance of measurement setup. Attempts to compare results from different research groups or different test settings require careful examination of test design details, and most often allow only limited paralleling conclusions to be drawn. This is because a combination of factors introducing variability will change from experiment to experiment.

Validation and calibration of instrumental equipment of course is crucial to reliable data acquisition. Experimental setup has to provide correct sampling conditions regarding temperature and pressure control of transfer lines, installation of dilution auxiliaries, application of proper dilution and aliquot ratios, adequate adsorbent and filter material and cross sections, and many more factors. An instructive article on background and practice of particulates measurement incl. aftertreatment issues with respect to the European legislative framework is given by Giechaskiel et al. (2012). A thorough evaluation of procedural requirements and errors for the determination of very small PM amounts emitted by a PFI and a GDI gasoline vehicle was performed by Jung et al. (2016) in a report for CRC. Readers interested in issues of emission data acquisition, quality control and processing may read through an article published by Sharma et al. (2012).

The type of test cycle and test protocol relates to the set of engine operation points to be covered in specified time steps and time frames, ambient conditions and system preconditioning to follow, and exhaust sampling and relative weighting of sampling points/windows while running through the test sequence. Gaseous emissions are monitored as concentration trace over time, or as integrated emission in case of bag or adsorption sampling. Particle number concentrations or particle size distributions are captured quasi-continuously (resolution depends on impactor/sizer/counter performance), while particle mass is an integrated value over the test run. In transient cycles, simulating dynamic real world driving with frequent intervals of ever changing speed/load conditions, quasi-stationary combustion cannot be attained and thus exhaust sampling takes place uniformly over the entire test period. In contrast, stationary cycles, on moving to the next speed/load engine setting, involve a stabilization phase until stable combustion is achieved, and sampling is performed during just a defined period of stable combustion. According to predefined test point weight factors, sample aliquots are introduced for each test point, contributing to an integrated parameter sampled by gasbag, adsorption or filter.

Since every engine speed/load condition exhibits specific combustion kinetics generating corresponding exhaust gas compositions, the choice of test sequence has a large influence on measurement results both in terms of quality and in terms of quantity. It may happen that one operation mode shows promising emissions but for another mode, the emission results are disastrous, or that advantages typically assigned to a fuel vanish upon application of markedly different test sequences. Kinetic modeling is an option to elucidate possible mechanistic causalities for such heterogeneous results. High transiency (dynamics) has dramatic effects as does engine starting (see results published by Roy et al. (2016) and Drozd et al. (2016) already cited) and consequently have been integrated in test sequences. The dramatic increase in particulate emissions of a 2002 transit bus equipped with DOC at acceleration and high-grade routes was documented in a publication of Sonntag et al. (2013). Armas et al. (2013) investigated cold and warm start events of diesel engines fueled with animal fat biodiesel, GTL and common diesel fuel. Somewhat expected results were higher cold start emissions and generally lower particle emissions of GTL and biodiesel, but differences between cold and warm start among tested fuels were equivocal in higher fossil diesel emissions at warm start and higher GTL/biodiesel NO_X emissions at cold start.

To discern cold and warm engine starting is an issue of system conditioning. System halts, cooling intervals, soak times at defined ambient temperatures all contribute to combustion characteristics and exhaust quality, as does temperature and humidity of intake air (see an investigation by Zhu et al. 2016 on GDI and PFI gasoline vehicles operated at +30 °C and -7 °C). Atmospheric pressure relates to engine efficiency, which is reflected in predetermined altitude ranges not to exceed with certification and approval testing. Preconditioning after changing fuel type may require 1000 km driving or running several non-evaluated cycles, but for convenience it may be acceptable to perform fuel comparison test runs by switching fuel valves and monitoring exhaust parameters until steady state is reached. He et al. (2011), when examining particulate matter emissions from mid-level gasoline blends of ethanol and isobutanol, find that instrumentation drift and unstable coolant and oil temperatures exert additional scatter or bias to particle size distributions. This leads to a dilemma in that prolonged warm-up periods, intended to provide stable measuring conditions, may turn out to pose new bias to measurement results.

There is no need to re-emphasize that fuel supply strategy and engine type are fundamental factors for exhaust emissions quality. Principles and characteristics of traditional, advanced and upcoming engine combustion processes, however, cannot be discussed here since this is not a key subject of this report and would cover undue space. After a run-through of the references on published emission testing investigations, the reader will recognize the huge number of different technical solutions that have been applied and that common subdivision of fuel components into "diesel-like" and "gasoline-like" gets more and more blurred due to advanced techniques of combustion control. To become familiar with engine and combustion principles we again recommend the review article by Bergthorson & Thomson (2014) already cited.

Concerning modes of fuel supply, some remarks will help categorize emission survey notes compiled in Section 7.2.3. Traditionally, one single type of liquid fuel is delivered and injected into the engine cylinder, intake air needed for combustion is used "as is", meaning that oxygen content and humidity are the only chemical parameters the modification of which might alter the combustion process. A serious modification of intake air composition is exerted by exhaust gas recirculation (EGR).

Since common fossil fuels *per se* are mixtures of many nonpolar combustible substances, blending with oxygenates or other additives does not introduce a different mode of fuel supply, as long as fuel comes as clear, homogenous liquid phase. Detergents or solubilizes will ensure homogeneity and prevent phase separation in case of larger polarity differences between fuel components, as in blending of methanol or ethanol in petroleum-based fuels. On the other hand, emulsion fuels comprised of a nonpolar main component with uniformly dispersed polar (hydrous) micelles have been introduced long ago such as water diesel. Significant improvements in emulsion stability can be achieved by preparing micro- and nano-emulsions. A review by Reham et al. (2015) describes preparation and properties of biofuel emulsions, Debnath et al. (2015) provide a comprehensive description on emulsion preparation, properties and effects on diesel engine emissions. Md Ishak et al. (2015) prepared and characterized a set of diesel-palm biodiesel-20% water microemulsions, Pereira et al. (2016) characterized *Babassu* biodiesel microemulsions by electrochemical impedance spectroscopy (EIS), phase diagram and IR. See also short articles on hydrous fuel application in MTZ worldwide (in English) by Dittmann et al. (2015) and Simon & Dörksen (2016) reporting NO_x and PM reductions, though PN increases at higher microemulsion water content. FAME isn't a promising part of a hydroemulsion fuel.

Minor amounts of water will influence combustion kinetics, but hydrous diesel fuel emulsions provide another mode of action by *flash-boiling*, i.e. instantaneous vaporization at injection into the hot cylinder. Flash vaporization enhances fuel spray atomization and thus cylinder charge homogenization. The principle is applicable to any combination of high-/low-boiling fuel components, such as diesel-water, diesel-gasoline, gasoline-LNG, biodiesel-DME and so on, as will be illustrated in Section 7.2.3. For an introduction to this topic as exemplified by gasoline DI engine testing see an article by Xu et al. (2015).

There may be reasons to keep fuel components with diverse properties separated instead of preparing blends or emulsions: (i) risk of phase separation or other incompatibilities affecting long-term stability, or (ii) injection strategies that perform on-demand dosage of a minor fuel component according to momentary engine status. The combustion process will respond to specific properties of fuel components entering the cylinder in varying amounts. An example was given already in Chapter 4 (Morganti et al. 2015 dealing with gasoline-ethanol "octane on demand"), and there is more to come in Section 7.2.3. This dual-fuel approach can be realized in two basic variants: injection of both fuel components via separate inlet valves, or injection of one component plus feed of the second by fumigation into intake air. Fumigated components enter the cylinder via plenum valve or nebulizing into the intake plenum airflow, which resembles the traditional carburetor technique. Again, water is a possible second component for direct injection or fumigation; see Böhm et al. (2016).

Dual-fuel operation requires twofold, separated fuel storage/feed configurations, which however poses no serious technical barrier. Common applications are engines operating on different fuels, such as neat plant oil engines employing fossil diesel for engine starting and turnoff, or gas engines that require diesel as pilot fuel. Selective catalytic reduction (SCR) exhaust aftertreatment comes with an additional reservoir for operational fluid too. While imposing extra costs for supplementary equipment, dual-fuel offers great potential benefits from in-time supply of distinct fuel compositions according to momentary engine operation requirements.

For a comparative overview on blend, emulsion and fumigation types of fuel supply and their effects on engine emission behavior, see Abedin et al. (2016).

The decision on which exhaust sampling point is to be chosen varies depending on the purpose of the investigation. Studying composition of raw exhaust or getting clues on combustion kinetics needs engine-out or diluter-out measurements, while overall vehicle performance must describe final (tailpipe) exhaust quality after having passed aftertreatment devices. Another experimental goal is to investigate effects of fuel variation on aftertreatment performance, whereby exhaust gas sampling consequently locates both upstream and downstream of respective modules.

Efficiency of three-way catalyst (TWC), SCR (selective catalytic reduction), DOC (diesel oxidation catalyst) and diesel particle filter (DPF) for exhaust pollutant reduction meanwhile is the ultimate backup for clean vehicles. To rely on late-stage refinement or elimination of inacceptable emissions requires sound maintenance of system performance and that vehicle users are aware of factors compromising the functionality of aftertreatment devices. Anyway, aging of catalysts will slowly deteriorate their efficiency, and consequently durability criteria have been implemented in vehicle approvals. Chen & Borken-Kleefeld (2016) in a recent fleet test investigation demonstrated the subtle worsening of NO_X emissions over a 15-year time course, depending on emission tier class of vehicles.

Beyond tailpipe sampling, "chase" sampling of exhaust plumes is a feasible way to collect data on exhaust quality; see applications reported by Pirjola et al. (2004), Zavala et al. (2009) and Hudda et al. (2013). This is a bridging towards ambient air pollution monitoring and research which was discussed earlier in this chapter, cf. respective investigations by Dallmann et al. (2014; exhaust plumes and highway tunnels) or Hudda & Fruin (2016; airport plume transect cruising) already cited.

Modification of engine control unit (ECU) – in order to attain optimum engine performance with respect to varying fuel properties it is necessary to adapt engine control units/modules (ECU/ECM) to fuel type. Such adaptations balance differences in fuel density, viscosity, compressibility, speed of sound, volatility, caloric value, cetane/octane number and oxygen content which affect fuel delivery (pumps, tubing, valves), injection timing, fuel vaporization/mixing and overall combustion progress. Efforts to adjust engine control according to fuel properties are indispensable if high proportions of non-hydrocarbon fuels (alcohols, ethers, biodiesel) are to be used, which requires specific information on fuel characteristics. However, ECU adaptions cannot cover a large matrix of fuel properties combined with any operational condition. Implementing fuel sensors capable of detecting fuel type actually delivered to the injector could only satisfy the demands of such a strategy. For a detailed investigation on engine control implications in biodiesel combustion, see an article by Hall et al. (2013).

Another course of action is the intentional employ of trade-offs for ECU adjustments. Settings would be optimized in favor of minimum emission of one exhaust component at the expense of another, while these latter worsened emissions are to be controlled by subsequent aftertreatment devices. However, ECU mapping is a demanding task, as besides emission parameter optimization, the tight interplay with other adjustments for e.g. fuel consumption, thermal efficiency, torque, noise, etc. has to be considered.

Therefore, in exhaust emission investigations it is likewise acceptable to keep ECU factory settings without adapting them to fuel type. To go even further, it may be advisable to intentionally maintain ECU settings unaltered in order to evaluate the effects of fuel type on combustion and to get clues on robustness of engine exhaust quality towards fuel compositional changes. Results thus obtained reflect everyday situations, since changes in fuel composition after refilling by customers/vehicle drivers will not entail garage stops in order to adjust engine control.

7.2.3 Influence of fuel composition on engine emissions

Preparation of blends – taking *refilling* as a keyword from the paragraph just above, we can continue with another keyword, *splash blending*, as an item that has generated remarkable debate along with interpretation of emission testing results (Griend 2013). Fuel refilling by vehicle users undoubtedly can be termed splash blending, which simply means pouring different fuels together into a receptacle (tank), not intending to apply additional mixing in order to get a homogeneous mixture. Such a simple procedural demand may even appear in test protocols, as exemplified in the "Protocol for the evaluation of effects of metallic fuel-additives on the emissions performance of vehicles" by EU Joint Research Center (JRC anonym. 2013). Guidelines in this protocol are in response to demands raised by fuel quality EU-Directive 2009/30/EC (FQD, European Parliament and the Council 2009), referring to limits for metalloid manganese contents. On p.14 of the JRC document, it says:

> "The candidate, metallic fuel-additive must only be splash blended into the fuel batch" [...]
> "Blending of the additive into the fuel must use methods that are practical in service" [...]
> "Artificial methods to ensure that the additive is homogeneously dispersed in the test fuel during the test program are not permitted."

More elaborate methods, to be applied by fuel blenders and filling stations in order to provide thoroughly mixed, homogenized fuel blends (e.g. E10, B7) of constant quality, are *sequential blending, ratio blending, hybrid blending* as a combination of these two, and *sidestream blending* (see e.g. an article by Gallehugh in Biodiesel Magazine 2008, "Biodiesel Blending Techniques Key to Quality Fuel", URL:

> http://www.biodieselmagazine.com/articles/2476/biodiesel-blending-techniques-key-to-quality-fuel/,

or a Technical Paper by FMC Solutions, "Biodiesel Blending Techniques Key to Quality Fuel" (2007), URL:

> http://info.smithmeter.com/literature/docs/tp0a015.pdf).

Application of proper blending practice is defined by a recent ASTM-method:

ASTM D 7794-14: Standard Practice for Blending Mid-Level Ethanol Fuel Blends for Flexible-Fuel Vehicles with Automotive Spark-Ignition Engines, current active version available at

> https://www.astm.org/Standards/D7794.htm
> http://www.techstreet.com/standards/astm-d7794-14?product_id=1890180.

Splash blending occurs in another, more distinct context as opposed to *match blending*. This refers to terms of delivery by fuel suppliers, which either allows near-customer (splash) blending by retailers making use of base fuel that itself fulfills specification demands already; or, situated at the refinery/fuel supplier, making use of base fuel that does not fully meet specification demands. In the latter case, starting with so-called *blendstock for oxygenate blending (BOB)*, base fuels receive additional bonus properties upon blending, after which specifications are met and fuels will be distributed to retailers/consumers; see an article by Rockstroh et al. (2016) on GTL naphtha for oxygenated gasoline.

It is evident that these types of blends, bearing deviating base fuel properties, will display slightly differing combustion characteristics; see an investigation on emission testing of ethanol-gasoline blends by Anderson et al. (2014). To achieve market specifications of final blends, fuel suppliers occasionally may vary minor BOB constituents without explicit notice, thereby introducing small modifications in chemical composition. Confusion or disputes on interpretation of assumed or observed fuel effects can be expected in such cases and require detailed fuel composition analyses.

Fuel quality – as a matter of quality control, fuels used for emission testing have to come with sound laboratory analyses of physicochemical properties, taking minor constituents or suspected trace impurities into account. Fossil base fuels, HVO/HEFA, XTL, DME, or lower alcohols purchased from commercial suppliers will be used "as received", with fuel properties and purity grades corresponding to industrial standards as documented in analysis certificates. Preparation of blends or any other modification of fuel starting material like addition of additives or aging entails re-examination of fuel parameters by corresponding analyses. Unfortunately, some authors fail to exact depict what fuel is applied, e.g. "biofuel" instead of "biodiesel" or, which butanol isomer out of four possible isomers has been used.

Speciation and quality of OME and biodiesel deserve special notice. As already mentioned in Section 0, synthesis of OME-type fuels still is being researched, requires strict control of reaction conditions, and may yield somewhat variable product qualities. Procedural demands for biodiesel preparation are much less complex than for OME, which allows researchers to prepare homemade fuel batches from conveniently available feedstock for their individual research purposes. Purity and speciation of homemade FAME are suspected to differ from certified commercial products, which introduces additional variance at biodiesel engine testing and uncertainty when attempting to compare published results.

Special attention has to be paid to traces of water within the fuel. Semi-polar or hydrophilic fuels like alcohols and biodiesel may contain residual water from production and are more susceptible to water uptake from ambient humidity than hydrocarbons. Investigations dealing with quantitative descriptions of aqueous fuels are given by Oliveira et al. (2012) on biodiesel and Pearson et al. (2014) on alcohol fuels. Diesel-biodiesel-blends are less prone to phase separation than gasoline-ethanol blends if significant amounts of water are present. Other possible adverse effects of fuel water traces are enhanced oxidative and microbial deterioration and corrosion of system parts.

As mentioned above in sections discussing fuel supply strategy topics and hydrous fuel emulsions, minor amounts of water have been recognized as having potential impacts on combustion kinetics and exhaust gas quality. The impetus for introducing water as chemical species arises from being a potential source of OH radicals and from its high enthalpy of vaporization with its charge cooling effect, thus depressing NO_x emissions. Control of water content is of special importance among commercially available ethanol blends, the alcohol component of which should come as roughly anhydrous ethanol but occasionally displays inferior quality grades with higher moisture content. The effects of water content variation in gasoline-ethanol blends on performance and emissions of SI or flex-fuel engines have been documented in publications by de Melo et al. (2011, 2012), Villela & Machado (2012), Stein et al. (2013), Kyriakides et al. (2013) and Daemme et al. (2016), the latter dealing with flex-fuel motorcycle emissions from Brazilian markets.

To minimize uncertainties associated with poorly defined or variable fuel composition, it may be necessary to perform dedicated studies with approved reference fuels like U.S. EPA's indolene. The

U.S. Code of Federal Regulations requirements for testing liquid fuels (diesel, gasoline, high-level ethanol-gasoline blends) are accessible online via

Diesel:
https://www.gpo.gov/fdsys/pkg/CFR-2014-title40-vol33/xml/CFR-2014-title40-vol33-sec1065-703.xml
http://www.ecfr.gov/cgi-bin/text-idx?SID=810d8437300ea37889d36a8d4c110e58&mc=true&node=se
40.37.1065_1703&rgn=div8

Gasoline:
https://www.gpo.gov/fdsys/pkg/CFR-2014-title40-vol33/xml/CFR-2014-title40-vol33-sec1065-710.xml
http://www.ecfr.gov/cgi-bin/text-idx?SID=810d8437300ea37889d36a8d4c110e58&mc=true&node=se
40.37.1065_1710&rgn=div8

Ethanol-gasoline blends (Exx):
https://www.gpo.gov/fdsys/pkg/CFR-2014-title40-vol33/xml/CFR-2014-title40-vol33-sec1065-725.xml
http://www.ecfr.gov/cgi-bin/text-idx?SID=810d8437300ea37889d36a8d4c110e58&mc=true&node=se
40.37.1065_1725&rgn=div8

See also a presentation on certification test fuel work by Machiele (2013).

The *Fuels for Advanced Combustion Engines (FACE)* program was initiated about ten years ago to establish reference fuels and corresponding test criteria for engine/vehicle approval and R&D purposes; see a status presentation by Zigler (2012) and a paper by Gallant et al. (2009) on physical and chemical properties of FACE diesels. Gieleciak & Fairbridge (2016) have accumulated emission data on FACE diesel fuels in the CRC Project AVFL-23 Final Report "Data Mining of FACE Diesel Fuels". In their executive summary they point out that: "… engine-out emissions and efficiency profiles depend strongly on engine operating parameters and to some extent on the chemical and physical properties of the diesel fuels" […], and that results from statistical modeling… " indicate that the number of secondary carbons in alkyl chains (determined by NMR) and aromatic carbons attached to alkyl groups are influential parameters on emissions".

Reference fuels by their proper definition of composition and physical determinants are especially valuable in fundamental studies on combustion and kinetics. Exemplary investigations in this respect were published by Kukkadapu & Sung (2015, autoignition study on two reference diesel fuels), Han (2013, synthetically designed diesel fuel properties and low-temperature diesel combustion), Fatouraie et al. (2012, Ignition and Combustion Properties of Ethanol-Indolene Blends in HCCI Engine) and Sarathy et al. (2016, ignition of FACE gasolines). To further reduce the number of chemical variables in fundamental investigations, fuel surrogate concepts exist which closely follow reference fuels as templates; see instructive articles by Anand et al. (2011), Reiter et al. (2015) and Elwardany et al. (2016). Comparative investigations on combustion behavior are provided by Javed et al. (2015) employing gasoline and n-heptane/iso-octane surrogate to study oxidation intermediates, and Kerschgens et al. (2015) studying three C_8-compounds (di-n-butylether, n-octanol, n-octane) in diesel engines.

The impact of trace metals and metalloid additives on fuel reactivity and deterioration has been discussed in Chapter 6 in terms of homogenous catalytic action in fuel liquids or liquid-solid interactions among metal surfaces of vehicle parts. Metal catalysis to be considered in conjunction with fuel combustion is a matter of homogenous gas phase reactions, as long as no reactive particles are involved. With enhancement of exhaust soot combustion (DPF regeneration) by metal traces, issues of heterogeneous gas-solid interactions and solid-state catalysis are covered. We will not go into mechanisms of metal catalysis but report on findings from application of fuel-borne metals/metalloids in engine combustion.

Metalloid fuel additives like methyl-cyclopentadienyl-manganese-tricarbonyl (MMT) and dicyclopentadienyl-iron (ferrocene) have been introduced as octane boosters, which points to their specific influence on radical chain processes during combustion. It is thus reasonable to assume effects on combustion particulate formation too, which in fact has led to application of trace metals as "soot- or smoke suppressants". To infer profound improvements in exhaust particulate quality from such an approach in any case is not legitimate, as the case of barium shows: Truex et al. (1980) investigated its effect on smoke opacity and PM and concluded that "reduced smoke opacity" to a great extent has to be attributed simply to an optical effect of white barium sulfate and possibly altered particle sizes instead of carbonaceous material depletion. "Smoke suppressant" thus must be termed a brash deception in this case.

A recent application of trace metal additivation pertains to their use as *fuel-borne catalyst (FBC)*, liberating metal ash particles which reside as active sites within the exhaust soot matrix. Though inter- ference of metals is effective as early as cylinder combustion takes place (affecting soot particle size, shape and speciation), the intended FBC effect is to lower filter regeneration temperature and reduce fuel late-injection to provide enhanced soot particle combustion in exhaust filter regeneration. Elemental species most often applied as FBC are oxidic cerium and iron, but other species like graphite oxide and aluminum oxide have been evaluated for combustion progress (Ooi et al. 2016). Similarly to metalloid octane boosters (MMT, ferrocene), *combustion catalysts* based e.g. on cerium dioxide fuel additives have been applied; see Kazerooni et al. (2016) for a characterization of microemulsion diesel additives and potentials to reduce HC, CO and soot emissions. An inherent drawback of fuel metal additive catalyst strategies is fouling and deposit forming from metal ashes in the engine and exhaust system which bears the risk of functionality deterioration; see for example Williams et al. (2014). Moreover, metal additives enhance oxidative attack on stored fuels and can reduce antioxidant effectivity, as has been described by Schober & Mittelbach (2005) for biodiesels.

Metal-laden particles escaping exhaust aftertreatment and thus entering the environment pose additional hazards and have been debated since metallic additives were used in fuels and engine oils. This is of special relevance in confined environments such as indoor workplaces and underground mines. As Bugarski (2012) points out in a symposium presentation, fuel borne catalysts intended to help filter regeneration should be prohibited for engines not equipped with such filters or in cases of doubtful filter performance. For manganese and MMT applications, Michalke & Fernsebner (2014) contributed to recent knowledge on environmental and toxic effects of manganese liberated by industrial and vehicular sources by a review on epidemiological studies, toxicological mechanisms, speciation and potential biomarkers for manganese exposure.

Two comprehensive reports on "Effects of Organometallic Additives on Gasoline Vehicles" (essentially MMT) and "Effect of Metallic Additives in Market Gasoline and Diesel" written by Broch & Hoekman (2015, 2016) have been published by CRC. Mudgal et al. (2013) performed a risk assessment and corresponding test methodology in response to demands of the EU fuel quality directive and ongoing debate on metallic fuel additives, with results documented in a report to the European Commission.

Since especially biofuels from their biogenic origin contain small amounts of trace elements, implications for exhaust soot speciation can be assumed. Indeed, as Ess et al. (2016a, 2016b) concluded from biofuel-diesel blend combustion measurements, soot reactivity from biofuels indicated some benefits possibly attributable to catalytic effects or at least favorable lattice/matrix disruptions.

Mühlbauer et al. (2016) indicated that soot reactivity is influenced by trace elements and oxygen content and that trace elements and ash constituents may originate from fuel, lubricating oil and engine wear.

Analytical chemists interested in instrumental trace element determination in biofuels are referred to publications by Sánchez et al. (2015) and Virgilio et al. (2015).

References

Aakko-Saksa, P.; Koponen, P.; Kihlman, J. et al. (2011): Biogasoline options for conventional spark-ignition cars. VTT Working Papers 187. Online: http://www.vtt.fi/publications/index.jsp

Abedin, M. J.; Imran, A.; Masjuki, H. H. et al. (2016): An overview on comparative engine performance and emission characteristics of different techniques involved in diesel engine as dual-fuel engine operation. Renewable and Sustainable Energy Reviews 60, pp. 306-316. DOI: 10.1016/j.rser.2016.01.118.

Adelhelm, C.; Niessner, R.; Poschl, U. et al. (2008): Analysis of large oxygenated and nitrated polycyclic aromatic hydrocarbons formed under simulated diesel engine exhaust conditions (by compound fingerprints with SPE/LC-API-MS). Analytical and Bioanalytical Chemistry 391 (7), pp. 2599-2608. DOI: 10.1007/s00216-008-2175-9.

Agarwal, A. K.; Gupta, T.; Dixit, N. et al. (2013): Assessment of toxic potential of primary and secondary particulates/aerosols from biodiesel vis-a-vis mineral diesel fuelled engine. Inhalation Toxicology 25 (6), pp. 325-332. DOI: 10.3109/08958378.2013.782515.

Aimanant, S. & Ziemann, P. J. (2013): Chemical Mechanisms of Aging of Aerosol Formed from the Reaction of n -Pentadecane with OH Radicals in the Presence of NO x. Aerosol Science and Technology 47 (9), pp. 979-990. DOI: 10.1080/02786826.2013.804621.

Alam, M. S.; Zeraati-Rezaei, S.; Stark, C. P. et al. (2016): The characterisation of diesel exhaust particles – composition, size distribution and partitioning. Faraday Discussions 189, pp. 69-84. DOI: 10.1039/c5fd00185d.

Alleman, T. (2013): Analysis of Ethanol Fuel Blends. SAE Int. J. Fuels Lubr. 6(3, pp. 870-876. DOI: 10.4271/2013-01-9071.

Alleman, T. L.; McCormick, R. L. & Yanowitz, J. (2015): Properties of Ethanol Fuel Blends Made with Natural Gasoline. Energy Fuels 29 (8), pp. 5095-5102. DOI: 10.1021/acs.energyfuels.5b00818.

Anand, K.; Ra, Y.; Reitz, R. D. et al. (2011): Surrogate Model Development for Fuels for Advanced Combustion Engines. Energy Fuels 25 (4), pp. 1474-1484. DOI: 10.1021/ef101719a.

Anderson, J. E.; Wallington, T. J.; Stein, R. A. et al. (2014): Issues with T50 and T90 as Match Criteria for Ethanol-Gasoline Blends. SAE Int. J. Fuels Lubr. 7 (3), pp. 1027-1040. DOI: 10.4271/2014-01-9080.

Andreau, K.; Leroux, M. & Bouharrour, A. (2012): Health and cellular impacts of air pollutants: from cytoprotection to cytotoxicity. Biochemistry Research International 2012, p. 493894. DOI: 10.1155/2012/493894.

Arcoumanis, C.; Bae, C.; Crookes, R. et al. (2008): The potential of di-methyl ether (DME) as an alternative fuel for compression-ignition engines: A review. Fuel 87 (7), pp. 1014-1030. DOI: 10.1016/j.fuel.2007.06.007.

Armas, O.; García-Contreras, R. & Ramos, Á. (2014): Pollutant emissions from New European Driving Cycle with ethanol and butanol diesel blends. Fuel Processing Technology 122, pp. 64-71. DOI: 10.1016/j.fuproc.2014.01.023.

Armas, O.; Gómez, A. & Ramos, Á. (2013): Comparative study of pollutant emissions from engine starting with animal fat biodiesel and GTL fuels. Fuel 113, pp. 560-570. DOI: 10.1016/j.fuel.2013.06.010.

Armbruster, H.; Stucki, S.; Olsson, E. et al. (2003): On-board conversion of alcohols to ethers for fumigation in compression ignition engines. Proceedings of the Institution of Mechanical Engineers, Part D: Journal of Automobile Engineering 217 (3), pp. 155-164. DOI: 10.1243/09544070360550444.

Asadauskas, S., Erhan, S.Z. (2001) Thin-Film Test to Investigate Liquid Oxypolymerization of Nonvolatile Analytes: Assessment of Vegetable Oils and Biodegradable Lubricant, JAOCS 78 (10), pp. 1029-1035

AuGOV – Australian Government. 2016. *Emission requirements*. Online: https://infrastructure.gov.au/roads/environment/files/Standards_for_Diesel_HDVs.pdf

Azizi, Z.; Rezaeimanesh, M.; Tohidian, T.; Rahimpour, M.R. (2014) Dimethyl ether: A review of technologies and production challenges. Chemical Engineering and Processing 82, pp. 150-172, DOI: 10.1016/j.cep.2014.06.007

Bahreini, R.; Middlebrook, A. M.; Gouw, J. A. de et al. (2012): Gasoline emissions dominate over diesel in formation of secondary organic aerosol mass. Geophys. Res. Lett. 39 (6), L06805, 6 pages. DOI: 10.1029/2011GL050718.

Baranton, S.; Uchida, H.; Tryk, D. A. et al. (2013): Hydrolyzed polyoxymethylenedimethylethers as liquid fuels for direct oxidation fuel cells. Electrochimica Acta 108, pp. 350-355. DOI: 10.1016/j.electacta.2013.06.138.

Beker, S. A.; da Silva, Y. P.; Bücker, F. et al. (2016): Effect of different concentrations of tert-butylhydroquinone (TBHQ) on microbial growth and chemical stability of soybean biodiesel during simulated storage. Fuel 184, pp. 701-707. DOI: 10.1016/j.fuel.2016.07.067.

Bergthorson, J. M. & Thomson, M. J. (2014): A review of the combustion and emissions properties of advanced transportation biofuels and their impact on existing and future engines. Renewable and Sustainable Energy Reviews 42, pp. 1393-1417. DOI: 10.1016/j.rser.2014.10.034.

bioliq®.2016. *Der bioliq®-Prozess*. Karlsruhe. Online: http://www.bioliq.de/55.php (19 August 2016)

Bin Sintang, M. D.; Rimaux, T.; van de Walle, D. et al. (2016): Oil structuring properties of monoglycerides and phytosterols mixtures. Eur. J. Lipid Sci. Technol. DOI: 10.1002/ejlt.201500517.

Bisig, C.; Roth, M.; Muller, L. et al. (2016): Hazard identification of exhausts from gasoline-ethanol fuel blends using a multi-cellular human lung model. Environmental Research 151, pp. 789-796. DOI: 10.1016/j.envres.2016.09.010.

Bisig, C.; Steiner, S.; Comte, P. et al. (2015): Biological Effects in Lung Cells In Vitro of Exhaust Aerosols from a Gasoline Passenger Car With and Without Particle Filter. Emiss. Control Sci. Technol. 1 (3), pp. 237-246. DOI: 10.1007/s40825-015-0019-6.

BMJV 2016 – Bundesministerium der Justiz und für Verbraucherschutz (Federal Ministry of Justice and consumer protection). 2016. *Energiesteuergesetz (EnergieStG)*. Online: https://www.gesetze-im-internet.de/bundesrecht/energiestg/gesamt.pdf (07.12.2016)

BMVIT – Bundesministerium für Verkehr, Innovation und Technology (Austrian Federal Ministry of Transportation, Innovation and Technology). 2016. *Aktionspaket zur Förderung der Elektromobilität.* Online: https://www.bmvit.gv.at/presse/aktuell/downloads/leichtfried/emobilpaket.pdf (14.12.2016)

Bocchi, C.; Bazzini, C.; Fontana, F. et al. (2016): Characterization of urban aerosol: seasonal variation of mutagenicity and genotoxicity of PM2.5, PM1 and semi-volatile organic compounds. Mutation Research 809, pp. 16-23. DOI: 10.1016/j.mrgentox.2016.07.007.

Böhm, M.; Mährle, W.; Bartelt, H.-C. et al. (2016): Functional Integration of Water Injection into the Gasoline Engine. MTZ Worldw. 77, pp. 36-41. DOI: 10.1007/s38313-015-0073-z.

Botella, L.; Bimbela, F.; Martin, L. et al. (2014): Oxidation stability of biodiesel fuels and blends using the Rancimat and PetroOXY methods. Effect of 4-allyl-2,6-dimethoxyphenol and catechol as biodiesel additives on oxidation stability. Frontiers in chemistry 2, p. 43. DOI: 10.3389/fchem.2014.00043.

Brines, M.; Dall'Osto, M.; Beddows, D. C. S. et al. (2015): Traffic and nucleation events as main sources of ultrafine particles in high-insolation developed world cities. Atmos. Chem. Phys. 15 (10), pp. 5929-5945. DOI: 10.5194/acp-15-5929-2015.

Britto, R. F. & Martins, C. A. (2015): Emission analysis of a Diesel Engine Operating in Diesel-Ethanol Dual-Fuel mode. Fuel 148, pp. 191-201. DOI: 10.1016/j.fuel.2015.01.008.

Broch, A. & Hoekman, K. (2016): Effect of Metallic Additives in Market Gasoline and Diesel, Final Report. CRC Report No. E-114-2. Online: https://crcao.org/reports/recentstudies2016/E-114-2/CRC%20E114-2_Final%20Report%20with%20cover%20page.pdf

Broch, A. & Hoekman, S. K. (2015): Effects of Organometallic Additives on Gasoline Vehicles: Analysis of Existing Literature, Final Report. CRC Report No. E-114. Online: https://crcao.org/reports/recentstudies2015/E-114/Final%20Report.pdf

Broustail, G.; Halter, F.; Seers, P. et al. (2012): Comparison of regulated and non-regulated pollutants with iso-octane/butanol and iso-octane/ethanol blends in a port-fuel injection Spark-Ignition engine. Fuel 94, pp. 251-261. DOI: 10.1016/j.fuel.2011.10.068.

Bücker, F.; Barbosa, C. S.; Quadros, P. D. et al. (2014): Fuel biodegradation and molecular characterization of microbial biofilms in stored diesel/biodiesel blend B10 and the effect of biocide. International Biodeterioration & Biodegradation 95, pp. 346-355. DOI: 10.1016/j.ibiod.2014.05.030.

Bugarski, A. (2012): Alternative Fuels as a Diesel Emissions Control Strategy. Diesel Aerosols and Gases in Underground Metal and Nonmetal Mines, 14th U.S. / North American Mine Ventilation Symposium Salt Lake City, Utah, June 17th, 2012. NIOSH, 2012. Online: https://www.cdc.gov/niosh/mining/UserFiles/workshops/dieselaerosols2012/BugarskiMVS02012AltFuels.pdf

Burger, J.; Siegert, M.; Ströfer, E. et al. (2010): Poly(oxymethylene) dimethyl ethers as components of tailored diesel fuel. Properties, synthesis and purification concepts. Fuel 89 (11), pp. 3315-3319. DOI: 10.1016/j.fuel.2010.05.014.

CAAFI – Commercial Aviation Alternative Fuels Initiative. 2010. *Fuel Readiness Level*. Online: http://www.caafi.org/information/pdf/frl_caafi_jan_2010_v16.pdf

Cacua, K.; Amell, A. & Cadavid, F. (2012): Effects of oxygen enriched air on the operation and performance of a diesel-biogas dual fuel engine. Biomass and Bioenergy 45, pp. 159-167. DOI: 10.1016/j.biombioe.2012.06.003.

Cadrazco, M.; Agudelo, J. R.; Orozco, L. Y. et al. (2016): Genotoxicity of Diesel Particulate Matter Emitted by Port-Injection of Hydrous Ethanol and N-Butanol. ASME 2016 Internal Combustion Engine 9 October 2016, V001T02A012, 7pages. Online: http://proceedings.asmedigitalcollection.asme.org/proceeding.aspx?articleid=2589981.

Camerlynck, S.; Chandler, J.; Hornby, B. et al. (2012): FAME Filterability. Understanding and Solutions. SAE Int. J. Fuels Lubr. 5 (3), pp. 968-976. DOI: 10.4271/2012-01-1589.

Canakci, M.; Ozsezen, A.N.; Alptekin, E.; Eyidogan, M.(2013) Impact of alcohol-gasoline fuel blends on the exhaust emission of an SI engine. Renewable Energy 52, pp. 111-117. DOI: 10.1016/j.renene.2012.09.062.

Cannella, W.; Fairbridge, C.; Gieleciak, R. et al. (2013): Advanced alternative and renewable diesel fuels: detailed characterization of physical and chemical properties. CRC Report No. AVFL-19-2. CRC. Online: https://crcao.org/reports/recentstudies2013/AVFL-19-2/CRC%20Project%20AVFL-19-2%20Final%20Report.pdf

Cardoso, C. C.; Celante, V. G.; Castro, E. V. R. de et al. (2014): Comparison of the properties of special biofuels from palm oil and its fractions synthesized with various alcohols. Fuel 135, pp. 406-412. DOI: 10.1016/j.fuel.2014.07.019.

Cavina, N.; Poggio, L.; Bedogni, F. et al. (2013): Benchmark Comparison of Commercially Available Systems for Particle Number Measurement. In: 11th International Conference on Engines & Vehicles, SEP. 15, 2013: SAE International, Warrendale, PA, United States (SAE Technical Paper Series).

Çay, Y.; Korkmaz, I.; Çiçek, A.; Kara, F. (2013): Prediction of engine performance and exhaust emissions for gasoline and methanol using artificial neural network. Energy 50, pp. 177-186. DOI: 10.1016/j.energy.2012.10.052.

Çelik, M.B.; Özdalyan, B.; Alkan, F. (2011): The use of pure methanol as fuel at high compression ratio in a single cylinder gasoline engine. Fuel 90 (4), pp. 1591-1598. DOI: 10.1016/j.fuel.2010.10.035.

Cha, J.; Kwon, S. & Park, S. (2012): Engine performance and exhaust emissions in stoichiometric combustion engines fuelled with dimethyl ether. Proceedings of the Institution of Mechanical Engineers, Part D: Journal of Automobile Engineering 226 (5), pp. 674-683. DOI: 10.1177/0954407011426032.

Chai, M. (2012): Thermal decomposition of methyl esters in biodiesel fuel: kinetics, mechanisms and products. Dissertation. University of Cincinnati. Online: https://etd.ohiolink.edu/letd.send_file?accession=ucin1342544227&disposition=attachment

Chan, T. W. (2015): The Impact of Isobutanol and Ethanol on Gasoline Fuel Properties and Black Carbon Emissions from Two Light-Duty Gasoline Vehicles. In: SAE 2015 World Congress & Exhibition, APR. 21, 2015: SAE International, Warrendale, PA, United States (SAE Technical Paper Series). Online: http://papers.sae.org/2015-01-1076/

Chen, Y. & Borken-Kleefeld, J. (2016): NOx Emissions from Diesel Passenger Cars Worsen with Age. Environmental Science & Technology 50 (7), pp. 3327-3332. DOI: 10.1021/acs.est.5b04704.

Chen, Z.; Liu, J.; Wu, Z. et al. (2013): Effects of port fuel injection (PFI) of n-butanol and EGR on combustion and emissions of a direct injection diesel engine. Energy Conversion and Management 76, pp. 725-731. DOI: 10.1016/j.enconman.2013.08.030.

Choe, E. & Min, D. B. (2006): Mechanisms and Factors for Edible Oil Oxidation. Comprehensive Reviews in Food Science and Food Safety 5, pp. 169-186.

Choi, B.-C.; Park, S. & Lee, Y.-J. (2013): De-NOx Performance of Combined System of Reforming Catalyst and LNT for a DME Engine. In: SAE/KSAE 2013 International Powertrains, Fuels & Lubricants Meeting, OCT. 21, 2013: SAE International, Warrendale, PA, United States (SAE Technical Paper Series).

Chotwichien, A.; Luengnaruemitchai, A. & Jai-In, S. (2009): Utilization of palm oil alkyl esters as an additive in ethanol-diesel and butanol-diesel blends. Fuel 88 (9), pp. 1618-1624. DOI: 10.1016/j.fuel.2009.02.047.

Chow, J. C.; Doraiswamy, P.; Watson, J. G. et al. (2008): Advances in Integrated and Continuous Measurements for Particle Mass and Chemical Composition. Journal of the Air & Waste Management Association 58 (2), pp. 141-163. DOI: 10.3155/1047-3289.58.2.141.

Christensen, E. & McCormick, R. L. (2014): Long-term storage stability of biodiesel and biodiesel blends. Fuel Processing Technology 128, pp. 339-348. DOI: 10.1016/j.fuproc.2014.07.045.

Christensen, E.; Yanowitz, J.; Ratcliff, M. et al. (2011): Renewable Oxygenate Blending Effects on Gasoline Properties. Energy Fuels 25, pp. 4723-4733. DOI: 10.1021/ef2010089.

Chuck, C. J.; Bannister, C. D.; Jenkins, R. W. et al. (2012): A comparison of analytical techniques and the products formed during the decomposition of biodiesel under accelerated conditions. Fuel 96, pp. 426-433. DOI: 10.1016/j.fuel.2012.01.043.

Chupka, G. M.; Yanowitz, J.; Chiu, G. et al. (2011): Effect of Saturated Monoglyceride Polymorphism on Low-Temperature Performance of Biodiesel. Energy Fuels 25 (1), pp. 398-405. DOI: 10.1021/ef1013743.

Cinar, C.; Can, Ö.; Sahin, F. et al. (2010): Effects of premixed diethyl ether (DEE) on combustion and exhaust emissions in a HCCI-DI diesel engine. Applied Thermal Engineering 30 (4), pp. 360-365. DOI: 10.1016/j.applthermaleng.2009.09.016.

Co, E. D. & Marangoni, A. G. (2012): Organogels: An Alternative Edible Oil-Structuring Method. J Am Oil Chem Soc 89 (5), pp. 749-780. DOI: 10.1007/s11746-012-2049-3.

Contino, F.; Masurier, J.-B.; Foucher, F. et al. (2014): CFD simulations using the TDAC method to model iso-octane combustion for a large range of ozone seeding and temperature conditions in a single cylinder HCCI engine. Fuel 137, pp. 179-184. DOI: 10.1016/j.fuel.2014.07.084.

Costa, Rodrigo C.; Sodré, José R. (2010): Hydrous ethanol vs. gasoline-ethanol blend. Engine performance and emissions. In: *Fuel* 89 (2), S. 287-293. DOI: 10.1016/j.fuel.2009.06.017.

Cummings, J. (2011): Effects of Fuel Ethanol Quality on Vehicle System Components. SAE Technical Paper 2011-01-1200. DOI: 10.4271/2011-01-1200.

Czerwinski, J. (2011-2014): Toxicity of Exhaust Gases and Particles from IC-Engines – International Activities Survey (EngToxIn). 1st/2nd/3rd/4th Information Report for IEA Implementing Agreement AMF, Annex XLII, international activities 2010/2011; 2011/2012; 2012/2013; 2013/2014. IEA AMF. Online: http://www.iea-amf.org/content/projects/map_projects/42

Daemme, L. C.; Penteado, R.; Corrêa, S. M. et al. (2016): Emissions of Criteria and Non-Criteria Pollutants by a Flex-Fuel Motorcycle. Journal of the Brazilian Chemical Society. DOI: 10.5935/0103-5053.20160111.

Dahmen, M. & Marquardt, W. (2015): A Novel Group Contribution Method for the Prediction of the Derived Cetane Number of Oxygenated Hydrocarbons. Energy Fuels 29 (9), pp. 5781-5801. DOI: 10.1021/acs.energyfuels.5b01032.

Dahmen, M., Marquardt, W. (2016): Model-Based Design of Tailor-Made Biofuels. Energy Fuels 30, pp. 1109-1134. DOI: 10.1021/acs.energyfuels.5b02674.

Dai, P.; Ge, Y.; Lin, Y. et al. (2013): Investigation on characteristics of exhaust and evaporative emissions from passenger cars fueled with gasoline/methanol blends. Fuel 113, pp. 10-16. DOI: 10.1016/j.fuel.2013.05.038.

Dallmann, T. R.; Onasch, T. B.; Kirchstetter, T. W. et al. (2014): Characterization of particulate matter emissions from on-road gasoline and diesel vehicles using a soot particle aerosol mass spectrometer. Atmos. Chem. Phys. 14 (14), pp. 7585-7599. DOI: 10.5194/acp-14-7585-2014.

Daniel, R.; Wang, C.; Xu, H. et al. (2012): Dual-Injection as a Knock Mitigation Strategy Using Pure Ethanol and Methanol. SAE Int. J. Fuels Lubr. 5 (2), pp. 772-784. DOI: 10.4271/2012-01-1152.

DBR 2016 – Die Bundesregierung (German Federal Government). 2016. *Weitere Steuervorteile für Elektroautos*. Online: https://www.bundesregierung.de/Content/DE/Artikel/2016/05/2016-05-18-elektromobilitaet.html (07.12.2016)

Debnath, B. K.; Saha, U. K. & Sahoo, N. (2015): A comprehensive review on the application of emulsions as an alternative fuel for diesel engines. Renewable and Sustainable Energy Reviews 42, pp. 196-211. DOI: 10.1016/j.rser.2014.10.023.

Deep, A.; Kumar, N.; Karnwal, A. et al. (2014): Assessment of the Performance and Emission Characteristics of 1-Octanol/Diesel Fuel Blends in a Water Cooled Compression Ignition Engine. In: SAE 2014 International Powertrain, Fuels & Lubricants Meeting, OCT. 20, 2014: SAE International, Warrendale, PA, United States (SAE Technical Paper Series). Online: http://papers.sae.org/2014-01-2830/

Delphi Corp. (2016): Worldwide Emission Standards. Heavy Duty and Off-Highway Vehicles. Online: http://delphi.com/docs/default-source/worldwide-emissions-standards/delphi-worldwide-emissions-standards-heavy-duty-off-highway-15-16.pdf

Destaillats, F. & Angers, P. (2005a): On the mechanisms of cyclic and bicyclic fatty acid monomer formation in heated edible oils. Eur. J. Lipid Sci. Technol. 107 (10), pp. 767-772. DOI: 10.1002/ejlt.200501159.

Destaillats, F. & Angers, P. (2005b): Thermally induced formation of conjugated isomers of linoleic acid. European Journal of Lipid Science and Technology 107 (3), pp. 167-172. DOI: 10.1002/ejlt.200401088.

Deutsch, D.; Oestreich, D.; Lautenschütz, P. et al. (2017): High Purity Oligomeric Oxymethylene Ethers as Diesel Fuels. Chem. Ing. Tech. 89(4), 486–489. DOI: 10.1002/cite.201600158.

Dittmann, P.; Dörksen, H.; Steiding, D. et al. (2015): Influence of Micro Emulsions on Diesel Engine Combustion. MTZ Worldw. 76, pp. 38-44. DOI: 10.1007/s38313-015-0008-8.

do N. Batista, L.; Da Silva, V. F.; Pissurno, É. C. G. et al. (2015): Formation of toxic hexanal, 2-heptenal and 2,4-decadienal during biodiesel storage and oxidation. Environ Chem Lett 13 (3), pp. 353-358. DOI: 10.1007/s10311-015-0511-9.

Dobereiner, G. E.; Erdogan, G.; Larsen, C. R. et al. (2014): A One-Pot Tandem Olefin Isomerization/Metathesis-Coupling (ISOMET) Reaction. ACS Catal. 4 (9), pp. 3069-3076. DOI: 10.1021/cs500889x.

Dodos, G. S.; Karonis, D.; Zannikos, F. et al. (2014): Assessment of the Oxidation Stability of Biodiesel Fuel using the Rancimat and the RSSOT methods. In: SAE 2014 International Powertrain, Fuels & Lubricants Meeting, OCT. 20, 2014: SAE International, Warrendale, PA, United States (SAE Technical Paper Series).

Donaldson, K.; Borm, P. J.; Castranova, V. et al. (2009): The limits of testing particle-mediated oxidative stress in vitro in predicting diverse pathologies; relevance for testing of nanoparticles. Particle and Fibre Toxicology 6, p. 13. DOI: 10.1186/1743-8977-6-13.

Drozd, G. T.; Zhao, Y.; Saliba, G. et al. (2016): Time Resolved Measurements of Speciated Tailpipe Emissions from Motor Vehicles: Trends with Emission Control Technology, Cold Start Effects, and Speciation. Environmental Science & Technology 50 (24), pp. 13592-13599. DOI: 10.1021/acs.est.6b04513.

Dubey, A. & Khaskin, E. (2016): Catalytic Ester Metathesis Reaction and Its Application to Transfer Hydrogenation of Esters. ACS Catalysis 6 (6), pp. 3998-4002. DOI: 10.1021/acscatal.6b00827.

Dwivedi, G. & Sharma, M. P. (2014): Impact of Antioxidant and Metals on Biodiesel Stability-A Review. J. Mater. Environ. Sci. 5, pp. 1412-1425.

EC – European Commission. 2010. *High-Level Expert Group on Key Enabling Technologies*. Brussels.

Echim, C.; Maes, J. & Greyt, W. D. (2012): Improvement of cold filter plugging point of biodiesel from alternative feedstocks. Fuel 93, pp. 642-648. DOI: 10.1016/j.fuel.2011.11.036.

Elghawi, U. M. & Mayouf, A. M. (2014): Carbonyl emissions generated by a (SI/HCCI) engine from winter grade commercial gasoline. Fuel 116, pp. 109-115. DOI: 10.1016/j.fuel.2013.07.124.

Elwardany, A. E.; Sazhin, S. S. & Im, H. G. (2016): A new formulation of physical surrogates of FACE A gasoline fuel based on heating and evaporation characteristics. Fuel 176, pp. 56-62. DOI: 10.1016/j.fuel.2016.02.041.

Enerkem (2017) Enerkem's Edmonton waste-to-biofuel facility receives registration approval from U.S. Environmental Protection Agency (EPA) to sell its ethanol under the U.S. Renewable Fuel Standard. Press release, online: http://enerkem.com/newsroom/?communique_id=122563

EPA – US Environmental Protection Agency. 2016. *EPA and DOT Finalize Greenhouse Gas and Fuel Efficiency Standards for Heavy-Duty Trucks*. Online: https://www.epa.gov/newsreleases/heavydutyaug162016 (07.12.2016)

Erkkil\am, Kimmo; Nylund, Nils-Olof; Hulkkonen, Tuomo; Tilli, Aki; Mikkonen, Seppo; Saikkonen, Pirjo et al. (2011): Emission performance of paraffinic HVO diesel fuel in heavy duty vehicles. In: SAE Technical Paper (2011-01-1966). DOI: 10.4271/2011-01-1966.

Ess, M. N.; Bladt, H.; Mühlbauer, W. et al. (2016a): Reactivity and structure of soot generated at varying biofuel content and engine operating parameters. Combustion and Flame 163, pp. 157-169. DOI: 10.1016/j.combustflame.2015.09.016.

Ess, M. N.; Bürger, M.; Mühlbauer, W. et al. (2016b): Examination of the Soot Reactivity of Different Diesel Fuels. MTZ Worldw. 77, pp. 66-71. DOI: 10.1007/s38313-015-0076-9.

European Parliament and the Council (1998): Directive 98/70/EC of the European Parliament and of the Council of 13 October 1998 relating to the quality of petrol and diesel fuels and amending Council Directive 93/12/EEC. Official Journal of the European Communities L350/58 ff., 28.12.1998. Online: http://eur-lex.europa.eu/legal-content/EN/TXT/PDF/?uri=CELEX:31998L0070&rid=1

European Parliament and the Council (2008): Directive 2008/50/EC of the European Parliament and of the Council of 21 May 2008 on ambient air quality and cleaner air for Europe. European Air Quality Directive. Official Journal of the European Union L 152/1 ff., 11.06.2008. Online: http://eur-lex.europa.eu/legal-content/EN/TXT/PDF/?uri=CELEX:32008L0050&rid=4

European Parliament and the Council (2009): Directive 2009/30/EC of the European Parliament and of the Council of 23 April 2009 amending Directive 98/70/EC as regards the specification of petrol, diesel and gas-oil and introducing a mechanism to monitor and reduce greenhouse gas emissions and amending Council Directive 1999/32/EC as regards the specification of fuel used by inland waterway vessels and repealing Directive 93/12/EEC. European Fuel Quality Directive (FQD). Official Journal of the European Union L 140/88ff., 5.6.2009. Online: http://eur-lex.europa.eu/legal-content/EN/TXT/PDF/?uri=CELEX:32009L0030&rid=12

F.O.Licht's World Ethanol and Biofuels Report, Volume 11, Issue 20, 2013.

Fachagentur Nachwachsende Rohstoffe e. V. (FNR). 2014. *Biokraftstoffe*. Rostock. Online: https://mediathek.fnr.de/media/downloadable/files/samples/b/r/brosch_biokraftstoffe_web.pdf (30 May 2016)

Fang, H. L. & McCormick, R. L. (2006): Spectroscopic Study of Biodiesel Degradation Pathways. In: Powertrain & Fluid Systems Conference 2006.

Fang, H. L.; Stehouwer, D. M. & Wang, J. C. (2003): Interaction Between Fuel Additive and Oil Contaminant: (II) Its Impact on Fuel Stability and Filter Plugging Mechanism. SAE Technical Paper 2003-01-3140, pp. 1-10.

Fang, Q.; Huang, Z.; Zhu, L.; Zhang, J-J; Xiao, J. (2011): Study on low nitrogen oxide and low smoke emissions in a heavy-duty engine fuelled with dimethyl ether. In: Proceedings of the Institution of Mechanical Engineers, Part D: Journal of Automobile Engineering 225 (6), S. 779-786. DOI: 10.1177/2041299110394513.

Farahani, M.; Pagé, D. & Turingia, M. P. (2011): Sedimentation in biodiesel and Ultra Low Sulfur Diesel Fuel blends. Fuel 90 (3), pp. 951-957. DOI: 10.1016/j.fuel.2010.10.046.

Fatouraie, M.; Keros, P. & Wooldridge, M. (2012): A Comparative Study of the Ignition and Combustion Properties of Ethanol-Indolene Blends During HCCI Operation of a Single Cylinder Engine. In: SAE 2012 World Congress & Exhibition, APR. 24, 2012: SAE International, Warrendale, PA, United States (SAE Technical Paper Series).

Feiling, A.; Münz, M. & Beidl, C. (2016): Potential of the Synthetic Fuel OME1b for the Soot-free Diesel Engine. ATZextra Worldw 21 (S11), pp. 16-21. DOI: 10.1007/s40111-015-0516-1.

Feng, Haojie; Sun, Ping; Liu, Junheng; Liu, Shenghua; Wang, Yumei (2016): Effect of PODE 3-8-Diesel Blended Fuel on Cumbustion and Emissions of Diesel Engine. In: Acta Petrolei Sinica, 816-822.

Fersner, A. S. & Galante-Fox, J. M. (2014): Biodiesel Feedstock and Contaminant Contributions to Diesel Fuel Filter Blocking. SAE Int. J. Fuels Lubr. 7 (3), pp. 783-791. DOI: 10.4271/2014-01-2723.

Fleisch, T. H.; Basu, A. & Sills, R. A. (2012): Introduction and advancement of a new clean global fuel. The status of DME developments in China and beyond. Journal of Natural Gas Science and Engineering 9, pp. 94-107. DOI: 10.1016/j.jngse.2012.05.012.

Flekiewicz, M.; Kubica, G. & Flekiewicz, B. (2014): The Analysis of Energy Conversion Efficiency in SI Engines for Selected Gaseous Fuels. SAE Technical Paper 2014-01-2692. DOI: 10.4271/2014-01-2692.

Flitsch, S.; Neu, P. M.; Schober, S. et al. (2014): Quantitation of Aging Products Formed in Biodiesel during the Rancimat Accelerated Oxidation Test. Energy Fuels 28 (9), pp. 5849-5856. DOI: 10.1021/ef501118r.

Fontaras, G.; Karavalakis, G.; Kousoulidou, M. et al. (2010): Effects of low concentration biodiesel blends application on modern passenger cars. Part 2: impact on carbonyl compound emissions. Environmental Pollution (Barking, Essex : 1987) 158 (7), pp. 2496-2503. DOI: 10.1016/j.envpol.2009.11.021.

Foong, T. M.; Morganti, K. J.; Brear, M. J. et al. (2014): The octane numbers of ethanol blended with gasoline and its surrogates. Fuel 115, pp. 727-739. DOI: 10.1016/j.fuel.2013.07.105.

Foucher, F.; Higelin, P.; Mounaïm-Rousselle, C. et al. (2013): Influence of ozone on the combustion of n-heptane in a HCCI engine. Proceedings of the Combustion Institute 34 (2), pp. 3005-3012. DOI: 10.1016/j.proci.2012.05.042.

Gallant, T.; Franz, J. A.; Alnajjar, M. S. et al. (2009): Fuels for Advanced Combustion Engines Research Diesel Fuels. Analysis of Physical and Chemical Properties. SAE Int. J. Fuels Lubr. 2 (2), pp. 262-272. DOI: 10.4271/2009-01-2769.

Gamble, J. F.; Nicolich, M. J. & Boffetta, P. (2012): Lung cancer and diesel exhaust: an updated critical review of the occupational epidemiology literature. Critical Reviews in Toxicology 42 (7), pp. 549-598. DOI: 10.3109/10408444.2012.690725.

Geng, Peng; Cao, Erming; Tan, Qinming; Wei, Lijiang (2017): Effects of alternative fuels on the combustion characteristics and emission products from diesel engines. A review. Renewable and Sustainable Energy Reviews 71, pp. 523-534. DOI: 10.1016/j.rser.2016.12.080.

Geng, P.; Yao, C.; Wei, L.; Liu, J.; Wang, Q.; Pan, W.; Wang, J. (2014): Reduction of PM emissions from a heavy-duty diesel engine with diesel/methanol dual fuel. Fuel 123, pp. 1-11. DOI: 10.1016/j.fuel.2014.01.056.

Geldenhuys, G.-L. (2014): Characterization of diesel emissions with respect to semi-volatile organic compounds in South African platinum mines and other confined environments. Dissertation. University of Pretoria, Pretoria, South Africa. Online: http://www.repository.up.ac.za/handle/2263/46248

Gentner, D. R.; Isaacman, G.; Worton, D. R. et al. (2012): Elucidating secondary organic aerosol from diesel and gasoline vehicles through detailed characterization of organic carbon emissions. Proceedings of the National Academy of Sciences of the United States of America 109 (45), pp. 18318-18323. DOI: 10.1073/pnas.1212272109.

Gerbrandt K, Chu P L, Simmonds A, Mullins K A, MacLean H L, Griffin W M, Saville B A. 2016. *Life cycle assessment of lignocellulosic ethanol: a review of key factors and methods affecting calculated GHG emissions and energy use.* Current Opinion in Biotechnology. 38: 63-70. DOI: 10.1016/j.copbio.2015.12.021

Gerlofs-Nijland, M. E.; Totlandsdal, A. I.; Tzamkiozis, T. et al. (2013): Cell toxicity and oxidative potential of engine exhaust particles: impact of using particulate filter or biodiesel fuel blend. Environmental Science & Technology 47 (11), pp. 5931-5938. DOI: 10.1021/es305330y.

Ghadikolaei, M. A. (2016): Effect of alcohol blend and fumigation on regulated and unregulated emissions of IC engines—A review. Renewable and Sustainable Energy Reviews 57, pp. 1440-1495. DOI: 10.1016/j.rser.2015.12.128.

Giakoumis, E. G.; Rakopoulos, C. D.; Dimaratos, A. M. et al. (2013): Exhaust emissions with ethanol or n-butanol diesel fuel blends during transient operation. A review. Renewable and Sustainable Energy Reviews 17, pp. 170-190. DOI: 10.1016/j.rser.2012.09.017.

Giechaskiel, B.; Mamakos, A.; Andersson, J. et al. (2012): Measurement of Automotive Nonvolatile Particle Number Emissions within the European Legislative Framework. A Review. Aerosol Science and Technology 46 (7), pp. 719-749. DOI: 10.1080/02786826.2012.661103.

Gieleciak, R. & Fairbridge, C. (2016): CRC Project AVFL-23: Data Mining of FACE Diesel Fuels, Final Report. NATURAL RESOURCES CANADA DIVISION REPORT CDEV-2016-0038-RE. Online:

https://crcao.org/reports/recentstudies2016/AVFL-23/CRC%20Project%20AVFL-23%20Final%20Report_Aug2016.pdf

Gill, D. W.; Ofner, H.; Stoewe, C. et al. (2014): An Investigation into the Effect of Fuel Injection System Improvements on the Injection and Combustion of DiMethyl Ether in a Diesel Cycle Engine. In: SAE 2014 International Powertrain, Fuels & Lubricants Meeting, OCT. 20, 2014: SAE International, Warrendale, PA, United States (SAE Technical Paper Series).

Gilmour, M. I.; Kim, Y. H. & Hays, M. D. (2015): Comparative chemistry and toxicity of diesel and biomass combustion emissions. Analytical and Bioanalytical Chemistry 407 (20), pp. 5869-5875. DOI: 10.1007/s00216-015-8797-9.

GLOBES – Israels Business Arena. 2016. *OECD: Israel fuel tax among highest.* Online: http://www.globes.co.il/en/article-oecd-israel-fuel-tax-among-highest-1001164041 (09.12.20016)

Godoi, R.; Polezer, G.; Borillo, G. C. et al. (2016): Influence on the oxidative potential of a heavy-duty engine particle emission due to selective catalytic reduction system and biodiesel blend. Science of the Total Environment 560-561, pp. 179-185. DOI: 10.1016/j.scitotenv.2016.04.018.

Goncalves, T. J.; Arnold, U.; Plessow, P. N. et al. (2017): Theoretical Investigation of the Acid Catalyzed Formation of Oxymethylene Dimethyl Ethers from Trioxane and Dimethoxymethane. ACS Catal. 7, 3615–3621. DOI: 10.1021/acscatal.7b00701.

Gordon, T. D.; Presto, A. A.; Nguyen, N. T. et al. (2014): Secondary organic aerosol production from diesel vehicle exhaust. Impact of aftertreatment, fuel chemistry and driving cycle. Atmos. Chem. Phys. 14 (9), pp. 4643-4659. DOI: 10.5194/acp-14-4643-2014.

Górski, K.; Sen, A. K.; Lotko, W. et al. (2013): Effects of ethyl-tert-butyl ether (ETBE) addition on the physicochemical properties of diesel oil and particulate matter and smoke emissions from diesel engines. Fuel 103, pp. 1138-1143. DOI: 10.1016/j.fuel.2012.09.004.

GOS – Government Offices of Sweden. 2016. *Fossil-free transport and travel: The Government's work to reduce the impact of transport on the climate.* Online: http://www.government.se/government-policy/environment/fossil-free-transport-and-travel-the-governments-work-to-reduce-the-impact-of-transport-on-the-climate/ (14.12.2016)

Griend, S.V. (2013): "Understanding the Emissions Benefits of Higher Ethanol Blends: EPA Modeling Fails to Tell the Whole Story"; Online: http://www.cleanfuelsdc.org/pubs/documents/CFDCVanderGriendWP_HR.pdf) (28.06.2018).

Grubbs, R. H. (2004): Olefin metathesis. Tetrahedron 60 (34), pp. 7117-7140. DOI: 10.1016/j.tet.2004.05.124.

Gu, X.; Li, G.; Jiang, X. et al. (2013): Experimental study on the performance of and emissions from a low-speed light-duty diesel engine fueled with n-butanol-diesel and isobutanol-diesel blends. Proceedings of the Institution of Mechanical Engineers, Part D: Journal of Automobile Engineering 227 (2), pp. 261-271. DOI: 10.1177/0954407012453231.

Guarieiro, A. L.; Santos, João V. da S.; Eiguren-Fernandez, A. et al. (2014): Redox activity and PAH content in size-classified nanoparticles emitted by a diesel engine fuelled with biodiesel and diesel blends. Fuel 116, pp. 490-497. DOI: 10.1016/j.fuel.2013.08.029.

Guillén, M. D. & Ruiz, A. (2004): Formation of hydroperoxy- and hydroxyalkenals during thermal oxidative degradation of sesame oil monitored by proton NMR. Eur. J. Lipid Sci. Technol. 106 (10), pp. 680-687. DOI: 10.1002/ejlt.200401026.

Guillén, M. D. & Ruiz, A. (2005): Monitoring the oxidation of unsaturated oils and formation of oxygenated aldehydes by proton NMR. Eur. J. Lipid Sci. Technol. 107 (1), pp. 36-47. DOI: 10.1002/ejlt.200401056.

Gubicza K, Nieves I U, Sagues W J, Barta Z, Shanmugam K T, Ingram L O. 2016. *Techno-economic analysis of ethanol production from sugarcane bagasse using a Liquefaction plus Simultaneous Saccharification and co-Fermentation process*. Bioresource Technology. 208: 42-48. DOI: 10.1016/j.biortech.2016.01.093

Gupta, T. & Singh, D. K. (2016): Organic Species Emitted as a Part of Combustion Residue: Fate and Transformation in the Ambient Air. Journal of Energy and Environmental Sustainability 1, pp. 10-18.

Hajbabaei, M.; Karavalakis, G.; Miller, J. W. et al. (2013): Impact of olefin content on criteria and toxic emissions from modern gasoline vehicles. Fuel 107, pp. 671-679. DOI: 10.1016/j.fuel.2012.12.031.

Hall, C. M.; van Alstine, D.; Kocher, L. et al. (2013): Closed-loop combustion control of biodiesel-diesel blends in premixed operating conditions enabled via high exhaust gas recirculation rates. Proc IMechE Part D: J Automobile Engineering 227, pp. 966-985. DOI: 10.1177/0954407013487647.

Haltenort, P.; Hackbarth, K.; Oestreich, D. et al. (2018): Heterogeneously catalyzed synthesis of oxymethylene dimethyl ethers (OME) from dimethyl ether and trioxane. Catal. Commun. 109, pp. 80–84. DOI: 10.1016/j.catcom.2018.02.013.

Han, I. H. & Csallany, A. S. (2009): Formation of Toxic α,β-Unsaturated 4-Hydroxy-Aldehydes in Thermally Oxidized Fatty Acid Methyl Esters. J Am Oil Chem Soc 86 (3), pp. 253-260. DOI: 10.1007/s11746-008-1343-6.

Han, S.B. (2017) A Study on Engine Performance and Emission Characteristics of Gasoline Spark Ignition Engine with Methanol and Ethanol Addition. Journal of Scientific and Engineering Research, 4(4) pp. 129-136

Han, M. (2013): The effects of synthetically designed diesel fuel properties – cetane number, aromatic content, distillation temperature, on low-temperature diesel combustion. Fuel 109, pp. 512-519. DOI: 10.1016/j.fuel.2013.03.039.

Hao, H.; Liu, F.; Liu, Z. et al. (2016): Compression ignition of low-octane gasoline: Life cycle energy consumption and greenhouse gas emissions. Applied Energy 181, pp. 391-398.

Happonen, M.; Heikkilä, J.; Aakko-Saksa, P. et al. (2013): Diesel exhaust emissions and particle hygroscopicity with HVO fuel-oxygenate blend. Fuel 103, pp. 380-386. DOI: 10.1016/j.fuel.2012.09.006.

HartEnergy (2014): International fuel quality standards and their implications for Australian standards. Online: https://www.environment.gov.au/system/files/resources/f83ff2dc-87a7-4cf9-ab24-6c25f2713f9e/files/international-feul-quality-standards.pdf

Härtl, M.; Gaukel, K.; Pélerin, D. et al. (2017): Oxymethylene Ether as Potentially CO2-neutral Fuel for Clean Diesel Engines Part 1: Engine Testing. MTZ Worldw. 78, pp. 52-59. DOI: 10.1007/s38313-016-0163-6.

Härtl, M.; Seidenspinner, P.; Jacob, E. et al. (2015): Oxygenate screening on a heavy-duty diesel engine and emission characteristics of highly oxygenated oxymethylene ether fuel. Fuel 153, pp. 328-335. DOI: 10.1016/j.fuel.2015.03.012.

Härtl, M.; Seidenspinner, P.; Wachtmeister, G. et al. (2014): Synthetic Diesel Fuel OME1 A Pathway Out of the Soot-NOx Trade-Off. MTZ Worldw. 75, pp. 48-53. DOI: 10.1007/s38313-014-0173-1.

Hasan, A. O.; Abu-jrai, A.; Al-Muhtaseb, A. H. et al. (2016): Formaldehyde, acetaldehyde and other aldehyde emissions from HCCI/SI gasoline engine equipped with prototype catalyst. Fuel 175, pp. 249-256. DOI: 10.1016/j.fuel.2016.02.005.

Hayati, I. N.; Man, Y. B. C.; Tan, C. P. et al. (2005): Monitoring peroxide value in oxidized emulsions by Fourier transform infrared spectroscopy. Eur. J. Lipid Sci. Technol. 107 (12), pp. 886-895. DOI: 10.1002/ejlt.200500241.

He, X.; Ireland, J. C.; Zigler, B. T. et al. (2011): Impacts of mid-level biofuel content in gasoline on SIDI engine-out and tailpipe particulate matter emissions. Conference Paper NREL/CP-5400-49311. NREL – National Renewable Energy Laboratory.

Hewu, W. & Longbao, Z. (2003): Performance of a direct injection diesel engine fuelled with a dimethyl ether/diesel blend. Proceedings of the Institution of Mechanical Engineers, Part D: Journal of Automobile Engineering 217 (9), pp. 819-824. DOI: 10.1177/095440700321700907.

Hill, L. (2013) Emissions Legislation Review. Proceedings of the conference Personalities of the Automotive Industry, Brașov, Online: http://www.unitbv.ro/Portals/32/conferinte/hill2013.pdf

Hoekman, S. K.; Broch, A.; Robbins, C. et al. (2011): Investigation of Biodiesel Chemistry, Carbon Footprint and Regional Fuel Quality. CRC Report No. AVFL-17a. CRC. Online: https://crcao.org/reports/recentstudies2011/AVFL-17a/CRC_AVFL-17a_Jan_2011.pdf

Hottenbach P, Brands T, Grünefeld G, Janssen A, Muether M, Pischinger S. 2010. Optical and Thermodynamic Investigations of Reference Fuels for Future Combustion Systems. SAE Int. J. Fuels Lubr. 3 (2): 819-838.

Hou, J.; Wen, Z.; Liu, Y. et al. (2014): Experimental study on the injection characteristics of dimethyl ether-biodiesel blends in a common-rail injection system. Proceedings of the Institution of Mechanical Engineers, Part D: Journal of Automobile Engineering 228 (3), pp. 263-271. DOI: 10.1177/0954407013507444.

Hrbek, J. 2016. Status report on thermal biomass gasification in countries participating in IEA Bioenergy Task 33. Vienna. Online: http://www.ieatask33.org/download.php?file=files/file/2016/Status%20report-corr_.pdf
https://infrastructure.gov.au/roads/environment/files/Emission_Standards_for_Petrol_Cars.pdf (09.12.2016)

Hsieh, P. Y. & Bruno, T. J. (2015): A perspective on the origin of lubricity in petroleum distillate motor fuels. Fuel Processing Technology 129, pp. 52-60. DOI: 10.1016/j.fuproc.2014.08.012.

Hu, E.; Hu, X.; Wang, X. et al. (2012): On the fundamental lubricity of 2,5-dimethylfuran as a synthetic engine fuel. Tribology International 55, pp. 119-125. DOI: 10.1016/j.triboint.2012.06.005.

Hudda, N.; Fruin, S.; Delfino, R. J. et al. (2013): Efficient determination of vehicle emission factors by fuel use category using on-road measurements: downward trends on Los Angeles freight corridor I-710. Atmospheric Chemistry and Physics 13 (1). DOI: 10.5194/acp-13-347-2013.

Hudda, N. & Fruin, S. A. (2016): International Airport Impacts to Air Quality: Size and Related Properties of Large Increases in Ultrafine Particle Number Concentrations. Environmental Science & Technology 50 (7), pp. 3362-3370. DOI: 10.1021/acs.est.5b05313.

Iannuzzi, S. E.; Barro, C.; Boulouchos, K. et al. (2016): Combustion behavior and soot formation/oxidation of oxygenated fuels in a cylindrical constant volume chamber. Fuel 167, pp. 49-59. DOI: 10.1016/j.fuel.2015.11.060.

ICCT – International Council of Clean Transportation. 2015. *Factsheet Brazil – Light Duty Vehicle Emission Standards*. Online: http://www.theicct.org/sites/default/files/info-tools/pvstds/Brazil_PVstds-facts_jan2015.pdf (18.11.2016)

ICCT – International Council of Clean Transportation. 2015. *Factsheet South Korea – Light Duty Vehicle Emission Standards*. Online: http://www.theicct.org/sites/default/files/info-tools/pvstds/Korea_PVstds-facts_jan2015.pdf (18.11.2016)

IEA (2014): The Potential and Challenges of Drop-in Biofuels. A Report by IEA Bioenergy Task 39. With assistance of Karatzos, S., McMillan, J.D., Saddler, J.N. IEA. Online: http://task39.sites.olt.ubc.ca/files/2014/01/Task-39-Drop-in-Biofuels-Report-FINAL-2-Oct-2014-ecopy.pdf.

IEA – International Energy Agency. 2016. *Global EV Outlook 2016*. Online: https://www.iea.org/publications/freepublications/publication/Global_EV_Outlook_2016.pdf (13.12.2016)

IEA-AMF – International Energy Agency – Advanced Motor Fuels. 2014. *Implementing Agreement. Strategic Plan 2015-2019*. Online: http://iea-amf.org/app/webroot/files/file/ExCo%20Meetings/ExCo%2047/25d%20Strategic%20Plan%20format.pdf

IFP – IFP Energies nouvelles. 2016. *Inauguration of the BioTfueL project demonstrator in Dunkirk: 2nd generation biodiesel and biokerosene production up and running*. Press Release (9. December 2016). Online: http://ifpenergiesnouvelles.com/News/Press-releases/Inauguration-of-the-BioTfueL-project-demonstrator-in-Dunkirk-2nd-generation-biodiesel-and-biokerosene-production-up-and-running (13.01.2017)

IMEP – Israel Ministry of Environmental Protection. 2015. *Reducing Pollution from Heavy Vehicle Fleets*. Online: http://www.sviva.gov.il/English/env_topics/AirQuality/PollutionFromTransportation/GovtMeasures/Pages/Reducing-Pollution-from-Heavy-Vehicle-Fleets.aspx (09.12.2016)

IMEP – Israel Ministry of Environmental Protection. 2016. *Government Measures to Reduce Vehicular Pollution*. Online:

http://www.sviva.gov.il/English/env_topics/AirQuality/PollutionFromTransportation/GovtMeasur es/Pages/default.aspx (09.12.2016)

Institute for Health and Consumer Protection, European Chemicals Bureau (2002): 1,3-Butadiene, Risk Assessment Report. EU. Online: http://echa.europa.eu/documents/10162/1f512549-5bf8-49a8-ba51-1cf67dc07b72.

Iyer, R. (2016): A review on the role of allylic and bis allylic positions in biodiesel fuel stability from reported lipid sources. Biofuels, pp. 1-12. DOI: 10.1080/17597269.2016.1236004.

Jain, S. & Sharma, M. P. (2010): Review of different test methods for the evaluation of stability of biodiesel. Renewable and Sustainable Energy Reviews 14 (7), pp. 1937-1947. DOI: 10.1016/j.rser.2010.04.011.

Jain, S. & Sharma, M. P. (2012): Correlation development between the oxidation and thermal stability of biodiesel. Fuel 102, pp. 354-358. DOI: 10.1016/j.fuel.2012.06.110.

Jakeria, M. R.; Fazal, M. A. & Haseeb, A. (2014): Influence of different factors on the stability of biodiesel. A review. Renewable and Sustainable Energy Reviews 30, pp. 154-163. DOI: 10.1016/j.rser.2013.09.024.

Jamsran, N. & Lim, O. T. (2014): A Study on the Autoignition Characteristics of DME-LPG Dual Fuel in HCCI Engine. 10th International Conference on Heat Transfer, Fluid Mechanics and Thermodynamics (HEFAT 2014). Online: http://www.repository.up.ac.za/handle/2263/44749

Janecek, D.; Rothamer, D. & Ghandhi, J. (2016): Investigation of cetane number and octane number correlation under homogenous-charge compression-ignition engine operation. Proceedings of the Combustion Institute (in press). DOI: 10.1016/j.proci.2016.08.015.

Janssen A, Muether M, Pischinger S, Kolbeck, A, Lamping M. 2010. *The Impact of Different Biofuel Components in Diesel Blends on Engine Efficiency and Emission Performance*. SAE Paper 2010-01-2119.

Jariyasopit, N.; McIntosh, M.; Zimmermann, K. et al. (2014): Novel nitro-PAH formation from heterogeneous reactions of PAHs with NO2, NO3/N2O5, and OH radicals: prediction, laboratory studies, and mutagenicity. Environmental Science & Technology 48 (1), pp. 412-419. DOI: 10.1021/es4043808.

Jeswani H K, Falano T, Azapagic A. 2016. *Life cycle environmental sustainability of lignocellulosic ethanol produced in integrated thermo-chemical biorefineries*. Biofuels, Bioproducts and Biorefining. 9(6): 661-676. DOI: 10.1002/bbb.1558

Javed, T.; Nasir, E. F.; Es-sebbar, E.-t. et al. (2015): A comparative study of the oxidation characteristics of two gasoline fuels and an n-heptane/iso-octane surrogate mixture. Fuel 140, pp. 201-208. DOI: 10.1016/j.fuel.2014.09.095.

Jeihouni, Y.; Ruhkamp, L. & Pischinger, S. (2013): Influence of the Combination of Fuel Properties for a DI-Diesel Engine Under Partly Homogeneous Combustion. In: SAE 2013 World Congress & Exhibition, APR. 16, 2013: SAE International, Warrendale, PA, United States (SAE Technical Paper Series).

Jenkins, R. W.; Sargeant, L. A.; Whiffin, F. M. et al. (2015): Cross-Metathesis of Microbial Oils for the Production of Advanced Biofuels and Chemicals. ACS Sustainable Chem. Eng. 3, pp. 1526-1535. DOI: 10.1021/acssuschemeng.5b00228.

Jin, W.; Su, S.; Wang, B. et al. (2016): Properties and cellular effects of particulate matter from direct emissions and ambient sources. Journal of Environmental Science and Health. Part A, Toxic/hazardous Substances & Environmental Engineering 51 (12), pp. 1075-1083. DOI: 10.1080/10934529.2016.1198632.

Jo, Y. S. (2016): More Effective Use of Fuel Octane in a Turbocharged Gasoline Engine: Combustion, Knock, Vehicle Impacts. Dissertation. Massachusetts Institute of Technology (MIT). Online: https://dspace.mit.edu/handle/1721.1/104246#files-area

Joeng, L.; Bakand, S. & Hayes, A. (2015): Diesel exhaust pollution. Chemical monitoring and cytotoxicity assessment. AIMS Environmental Science 2 (3), pp. 718-736. DOI: 10.3934/environsci.2015.3.718.

Jorgensen, S. W.; Cannella, W.; Bays, T. et al. (2015): Detailed characterization of physical and chemical properties of cellulosic gasoline stocks. CRC Report No. AVFL-19a. CRC. Online: https://crcao.org/reports/recentstudies2015/AVFL-19a/AVFL-19a%20Final%20Report_12-17-2015.pdf

Jose, T. K. & Anand, K. (2016): Effects of biodiesel composition on its long term storage stability. Fuel 177, pp. 190-196. DOI: 10.1016/j.fuel.2016.03.007.

JRC anonym. (2013): Protocol for the evaluation of effects of metallic fuel- additives on the emissions performance of vehicles. EU – Joint Research Center JRC. Online: https://ec.europa.eu/clima/sites/clima/files/transport/fuel/docs/fuel_metallic_additive_protocol_en.pdf

Jung, H.; Johnson, K. C.; Rusell, R. L. et al. (2016): Very Low PM Mass Measurements. Final Report CRC Project No. E-99. CRC. Online: https://crcao.org/reports/recentstudies2015/E-99/CRC_ARB%20Final%20Report_E-99_v17.pdf

Kaminski, H.; Kuhlbusch, T.; Rath, S. et al. (2013): Comparability of mobility particle sizers and diffusion chargers. Journal of Aerosol Science 57, pp. 156-178. DOI: 10.1016/j.jaerosci.2012.10.008.

Karavalakis, G.; Short, D.; Vu, D. et al. (2015a): A Complete Assessment of the Emissions Performance of Ethanol Blends and Iso-Butanol Blends from a Fleet of Nine PFI and GDI Vehicles. SAE Int. J. Fuels Lubr. 8 (2). DOI: 10.4271/2015-01-0957.

Karavalakis, G.; Short, D.; Vu, D. et al. (2015b): Evaluating the Effects of Aromatics Content in Gasoline on Gaseous and Particulate Matter Emissions from SI-PFI and SIDI Vehicles. Environmental Science & Technology 49, pp. 7021-7031. DOI: 10.1021/es5061726.

Karavalakis, G.; Jiang, Yu; Yang, Jiacheng; Durbin, T.; Nuottimäki, J.; Lehto, K. (2016): Emissions and Fuel Economy Evaluation from Two Current Technology Heavy-Duty Trucks Operated on HVO and FAME Blends. In: SAE Int. J. Fuels Lubr. 9 (1), S. 177-190. DOI: 10.4271/2016-01-0876.

Kazerooni, H.; Rouhi, A.; Khodadadi, A. A. et al. (2016): Effects of Combustion Catalyst Dispersed by a Novel Microemulsion Method as Fuel Additive on Diesel Engine Emissions, Performance, and Characteristics. Energy Fuels 30 (4), pp. 3392-3402. DOI: 10.1021/acs.energyfuels.6b00004.

Kelly, G. 2016. *Avenues to sustainable road transport energy in New Zealand*. International Journal of Sustainable Transportation. 10(6): 505-516. DOI: 10.1080/15568318.2015.1011795

Kerschgens, B.; Cai, L.; Pitsch, H. et al. (2015): Di-n-buthylether, n-octanol, and n-octane as fuel candidates for diesel engine combustion. Combustion and Flame 163, pp. 66-78. DOI: 10.1016/j.combustflame.2015.09.001.

Kivevele, T. & Huan, Z. (2015a): Influence of metal contaminants and antioxidant additives on storage stability of biodiesel produced from non-edible oils of Eastern Africa origin (Croton megalocarpus and Moringa oleifera oils). Fuel 158, pp. 530-537. DOI: 10.1016/j.fuel.2015.05.047.

Kivevele, T. & Huan, Z. (2015b): Review of the stability of biodiesel produced from less common vegetable oils of African origin. S. Afr. J. Sci 111 (9/10). DOI: 10.17159/sajs.2015/20140434.

Kleinová, A.; Cvengrošová, Z. & Cvengroš, J. (2013): Standard methyl esters from used frying oils. Fuel 109, pp. 588-596. DOI: 10.1016/j.fuel.2013.03.028.

Knothe, G. & Dunn, R. O. (2003): Dependence of oil stability index of fatty compounds on their structure and concentration and presence of metals. J Amer Oil Chem Soc 80 (10), pp. 1021-1026. DOI: 10.1007/s11746-003-0814-x.

Knothe, G. (2006a): Analysis of oxidized biodiesel by1H-NMR and effect of contact area with air. Eur. J. Lipid Sci. Technol. 108 (6), pp. 493-500. DOI: 10.1002/ejlt.200500345.

Knothe, G. (2006b): Analyzing Biodiesel: Standards and Other Methods. Review. JAOCS 83(10), pp. 823-833.

Knothe, G. (2008): "Designer" Biodiesel: Optimizing Fatty Ester Composition to Improve Fuel Properties. Energy Fuels 22 (2), pp. 1358-1364. DOI: 10.1021/ef700639e.

Knothe, G. (2014): A comprehensive evaluation of the cetane numbers of fatty acid methyl esters. Fuel 119, pp. 6-13. DOI: 10.1016/j.fuel.2013.11.020.

Knothe, G.; Cermak, S. C. & Evangelista, R. L. (2012): Methyl esters from vegetable oils with hydroxy fatty acids: Comparison of lesquerella and castor methyl esters. Fuel 96, pp. 535-540. DOI: 10.1016/j.fuel.2012.01.012.

Knothe, G.; Matheaus, A. C. & Ryan, T. W. (2003): Cetane numbers of branched and straight-chain fatty esters determined in an ignition quality tester. Fuel 82 (8), pp. 971-975. DOI: 10.1016/S0016-2361(02)00382-4.

Koivisto, E.; Ladommatos, N. & Gold, M. (2015a): Systematic study of the effect of the hydroxyl functional group in alcohol molecules on compression ignition and exhaust gas emissions. Fuel 153, pp. 650-663. DOI: 10.1016/j.fuel.2015.03.042.

Koivisto, E.; Ladommatos, N. & Gold, M. (2015b): The influence of various oxygenated functional groups in carbonyl and ether compounds on compression ignition and exhaust gas emissions. Fuel 159, pp. 697-711. DOI: 10.1016/j.fuel.2015.07.018.

Koivisto, E.; Ladommatos, N. & Gold, M. (2016): Compression ignition and pollutant emissions of large alkylbenzenes. Fuel 172, pp. 200-208. DOI: 10.1016/j.fuel.2016.01.025.

Konno, M. & Chen, Z. (2005): Ignition Mechanisms of HCCI Combustion Process Fueled With Methane/DME Composite Fuel. SAE Technical Paper 2005-01-0182. DOI: 10.4271/2005-01-0182.

Kousoulidou, M.; Fontaras, G.; Ntziachristos, L. et al. (2013): Use of portable emissions measurement system (PEMS) for the development and validation of passenger car emission factors. Atmospheric Environment 64, pp. 329-338. DOI: 10.1016/j.atmosenv.2012.09.062.

Kukkadapu, G. & Sung, C.-J. (2015): Autoignition study of ULSD#2 and FD9A diesel blends. Combustion and Flame 166, pp. 45-54. DOI: 10.1016/j.combustflame.2015.12.022.

Kuronen M, Mikkonen S. (2007) Hydrotreated vegetable oil as fuel for heavy duty diesel engines. SAE 2007-01-4031

Kyriakides, A.; Dimas, V.; Lymperopoulou, E. et al. (2013): Evaluation of gasoline-ethanol-water ternary mixtures used as a fuel for an Otto engine. Fuel 108, pp. 208-215. DOI: 10.1016/j.fuel.2013.02.035.

Landälv, I. 2016. *METHANOL Production from Biomass*. Presentation Green Pilot Kickoff Seminar (16 June 2016). Online: http://www.greenpilot.marinemethanol.com/pages/news/160601/2-02_20160616_Ingvar%20LandalvLTU.pdf

Lane, J. (2016): Norway to boost biofuels blend level to 20 percent. (10 December 2016). Online: http://www.biofuelsdigest.com/bdigest/2016/12/10/norway-to-bost-biofuels-blend-levels-to-20-percent/.

Langenhoven, J. (2014): The effect of humidity and soluble water content on the lubricty testing of a n-hexadecane and palmitic acid test fluid. Dissertation. Pretoria, South Africa. Online: http://www.repository.up.ac.za/handle/2263/46242

Langhorst, T.; Toedter, O.; Koch, T. et al. (2018): Investigations on Spark and Corona Ignition of Oxymethylene Ether-1 and Dimethyl Carbonate Blends with Gasoline by High-Speed Evaluation of OH* Chemiluminescence. SAE Int. J. Fuels Lubr. 11(1), pp. 1–15. DOI: 10.4271/04-11-01-0001. Open Access: https://saemobilus.sae.org/content/04-11-01-0001/

Lapuerta, M.; García-Contreras, R.; Campos-Fernández, J. et al. (2010): Stability, Lubricity, Viscosity, and Cold-Flow Properties of Alcohol–Diesel Blends. Energy Fuels 24 (8), pp. 4497-4502. DOI: 10.1021/ef100498u.

Lapuerta, M.; Martos, F. J. & Cárdenas, M. D. (2005): Determination of light extinction efficiency of diesel soot from smoke opacity measurements. Meas. Sci. Technol. 16 (10), pp. 2048-2055. DOI: 10.1088/0957-0233/16/10/021.

Lapuerta, M.; Villajos, M.; Agudelo, J. R. et al. (2011): Key properties and blending strategies of hydrotreated vegetable oil as biofuel for diesel engines. Fuel Processing Technology 92 (12), pp. 2406-2411. DOI: 10.1016/j.fuproc.2011.09.003.

Latza, U.; Gerdes, S. & Baur, X. (2009): Effects of nitrogen dioxide on human health: systematic review of experimental and epidemiological studies conducted between 2002 and 2006.

International Journal of Hygiene and Environmental Health 212 (3), pp. 271-287. DOI: 10.1016/j.ijheh.2008.06.003.

Lautenschütz, L. P. (2015): Neue Erkenntnisse in der Syntheseoptimierung oligomerer Oxymethylendimethylether aus Dimethoxymethan und Trioxan. Dissertation (in German), Heidelberg.

Lautenschütz, L.; Oestreich, D.; Seidenspinner, P. et al. (2016): Physico-chemical properties and fuel characteristics of oxymethylene dialkyl ethers. Fuel 173, pp. 129-137. DOI: 10.1016/j.fuel.2016.01.060.

Lautenschütz, L.; Oestreich, D.; Seidenspinner, P. et al. (2017): Corrigendum to "Physico-chemical properties and fuel characteristics of oxymethylene dialkyl ethers" [Fuel 173 (2016) 129-137]". Fuel 209, p. 812. DOI: 10.1016/j.fuel.2017.07.083.

Lautenschütz, L.; Oestreich, D.; Haltenort, P. et al. (2017): Efficient synthesis of oxymethylene dimethyl ethers (OME) from dimethoxymethane and trioxane over zeolites. Fuel Process. Technol. 165, pp. 27–33. DOI: 10.1016/j.fuproc.2017.05.005.

Lee, D.; Miller, A.; Kittelson, D. et al. (2006): Characterization of metal-bearing diesel nanoparticles using single-particle mass spectrometry. Journal of Aerosol Science 37 (1), pp. 88-110. DOI: 10.1016/j.jaerosci.2005.04.006.

Lee, S.; Oh, S. & Choi, Y. (2009): Performance and emission characteristics of an SI engine operated with DME blended LPG fuel. Fuel 88 (6), pp. 1009-1015. DOI: 10.1016/j.fuel.2008.12.016.

Lee, S.; Oh, S.; Choi, Y. et al. (2011a): Effect of n-Butane and propane on performance and emission characteristics of an SI engine operated with DME-blended LPG fuel. Fuel 90 (4), pp. 1674-1680. DOI: 10.1016/j.fuel.2010.11.040.

Lee, S.; Oh, S.; Choi, Y. et al. (2011b): Performance and emission characteristics of a CI engine operated with n-Butane blended DME fuel. Applied Thermal Engineering 31 (11-12), pp. 1929-1935. DOI: 10.1016/j.applthermaleng.2011.02.039.

Li, G.; Zhou, L.; Liu, S. et al. (2007): Experimental study on vapour pressure of dimethyl ether blended in diesel oil. Proceedings of the Institution of Mechanical Engineers, Part D: Journal of Automobile Engineering 221 (7), pp. 889-892. DOI: 10.1243/09544070JAUTO477.

Li, D.; Gao, Y.; Liu, S.; Ma, Z.; Wei, Y. (2016): Effect of polyoxymethylene dimethyl ethers addition on spray and atomization characteristics using a common rail diesel injection system. *Fuel* 186, pp. 235-247. DOI: 10.1016/j.fuel.2016.08.082.

Liati, A.; Dimopoulos Eggenschwiler, P.; Schreiber, D. et al. (2013): Variations in diesel soot reactivity along the exhaust after-treatment system, based on the morphology and nanostructure of primary soot particles. Combustion and Flame 160 (3), pp. 671-681. DOI: 10.1016/j.combustflame.2012.10.024.

Liati, A.; Schreiber, D.; Dimopoulos Eggenschwiler, P. et al. (2016): Electron microscopic characterization of soot particulate matter emitted by modern direct injection gasoline engines. Combustion and Flame 166, pp. 307-315. DOI: 10.1016/j.combustflame.2016.01.031.

Liu, J.; Wang, H.; Li, Y. et al. (2016): Effects of diesel/PODE (polyoxymethylene dimethyl ethers) blends on combustion and emission characteristics in a heavy duty diesel engine. Fuel 177, pp. 206-216. DOI: 10.1016/j.fuel.2016.03.019.

Liu, H.; Wang, Z.; Wang, J.; He, X.; Zheng, Y.; Tang, Q.; Wang, J.(2015): Performance, combustion and emission characteristics of a diesel engine fueled with polyoxymethylene dimethyl ethers (PODE3-4)/ diesel blends. In: Energy 88, S. 793-800. DOI: 10.1016/j.energy.2015.05.088.

Liu, P.; Kerr, B. J.; Chen, C. et al. (2014): Methods to create thermally oxidized lipids and comparison of analytical procedures to characterize peroxidation. Journal of animal science 92 (7), pp. 2950-2959. DOI: 10.2527/jas.2012-5708.

Liu, T.; Wang, X.; Deng, W. et al. (2015): Secondary organic aerosol formation from photochemical aging of light-duty gasoline vehicle exhausts in a smog chamber. Atmos. Chem. Phys. 15 (15), pp. 9049-9062. DOI: 10.5194/acp-15-9049-2015.

López, A. F.; Cadrazco, M.; Agudelo, A. F. et al. (2015): Impact of n-butanol and hydrous ethanol fumigation on the performance and pollutant emissions of an automotive diesel engine. Fuel 153, pp. 483-491. DOI: 10.1016/j.fuel.2015.03.022.

Louis, C.; Liu, Y.; Tassel, P. et al. (2016): PAH, BTEX, carbonyl compound, black-carbon, NO2 and ultrafine particle dynamometer bench emissions for Euro 4 and Euro 5 diesel and gasoline passenger cars. Atmospheric Environment 141, pp. 80-95. DOI: 10.1016/j.atmosenv.2016.06.055.

Lu, M. & Chai, M. (2011): Experimental Investigation of the Oxidation of Methyl Oleate: One of the Major Biodiesel Fuel Components. In Arno de Klerk, David L. King (Eds.): Synthetic liquids production and refining. Boston Meeting 2010, vol. 1084. Washington, DC: American Chemical Society (ACS Symposium Series, 1084), pp. 289-312.

Lu, X.-C.; Ma, J.-J.; Ji, L.-B. et al. (2008): Effects of premixed n-heptane from the intake port on the combustion characteristics and emissions of biodiesel-fuelled engines. Proceedings of the Institution of Mechanical Engineers, Part D: Journal of Automobile Engineering 222 (6), pp. 1001-1009. DOI: 10.1243/09544070JAUTO743.

Luecke, J. & McCormick, R. L. (2014): Electrical Conductivity and pH e Response of Fuel Ethanol Contaminants. Energy Fuels 28 (8), pp. 5222-5228. DOI: 10.1021/ef5013038.

Lumpp, B.; Rothe, D.; Pastötter, C. et al. (2011): Oxymethylene ethers as diesel fuel additives of the future. MTZ Worldw. 72 (3). pp. 34-38. DOI: 10.1365/s38313-011-0027-z.

Machiele, P. (2013): Vehicle Certification Test Fuel and Ethanol Flex Fuel Quality. Biomass 2013 Conference, August 1. U.S. EPA, 2013. Online: http://energy.gov/sites/prod/files/2014/05/f15/b13_machiele_2-b.pdf

Madden, M. C. (2016): A paler shade of green? The toxicology of biodiesel emissions: Recent findings from studies with this alternative fuel. Biochimica et Biophysica Acta (BBA) – General Subjects Vol 1860 (12), pp. 2856-2862.

Majer S., Gröngröft A. 2010. *Ökologische und ökonomische Bewertung der Produktion von Biomethanol für die Biodieselbereitstellung.* Union zur Förderung von Öl- und Proteinpflanzen (Hrsg.)

Mamakos, A.; Bonnel, P.; Perujo, A. et al. (2013): Assessment of portable emission measurement systems (PEMS) for heavy-duty diesel engines with respect to particulate matter. Journal of Aerosol Science 57, pp. 54-70. DOI: 10.1016/j.jaerosci.2012.10.004.

Mamakos, A.; Martini, G. & Manfredi, U. (2013): Assessment of the legislated particle number measurement procedure for Euro 5 and Euro 6 compliant diesel passenger cars under regulated and unregulated conditions. Journal of Aerosol Science 55, pp. 31-47. DOI: 10.1016/j.jaerosci.2012.07.012.

Manzetti, S. (2012): Are polycyclic aromatic hydrocarbons from fossil emissions potential hormone-analogue sources for modern man? Letter to the Editor. Pathophysiology 19, pp. 65-67.

Manzetti, S. & Andersen, O. (2015): A review of emission products from bioethanol and its blends with gasoline. Background for new guidelines for emission control. Fuel 140, pp. 293-301. DOI: 10.1016/j.fuel.2014.09.101.

Manzetti, S. & Andersen, O. (2016): Biochemical and physiological effects from exhaust emissions. A review of the relevant literature. Pathophysiology: The Official Journal of the International Society for Pathophysiology 23 (4), pp. 285-293. DOI: 10.1016/j.pathophys.2016.10.002.

Maricq, M. M. (2011): Physical and chemical comparison of soot in hydrocarbon and biodiesel fuel diffusion flames: A study of model and commercial fuels. Combustion and Flame 158, pp. 105-116. DOI: 10.1016/j.combustflame.2010.07.022.

Maricq, M. M. (2014): Examining the Relationship Between Black Carbon and Soot in Flames and Engine Exhaust. Aerosol Science and Technology 48, pp. 620-629. DOI: 10.1080/02786826.2014.904961.

Maus, W.; Jacob, E. 2014. *Synthetic Fuels—OME1: A Potentially Sustainable Diesel Fuel*. Online: http://www.emitec.com/fileadmin/user_upload/Bibliothek/Vortraege/140217_Maus_Jacob_LVK_Wien_englisch.pdf

Mayo, M. P. & Boehman, A. L. (2015): Ignition Behavior of Biodiesel and Diesel under Reduced Oxygen Atmospheres. Energy Fuels 29 (10), pp. 6793-6803. DOI: 10.1021/acs.energyfuels.5b01439.

Mc Ilrath, S. P.; Dash, B. P.; Topinka, M. J. et al. (2015): Reforming Biodiesel Fuels via Metathesis with Light Olefins. Current Green Chemistry 2, pp. 392-398.

McClellan, R. O.; Hesterberg, T. W. & Wall, J. C. (2012): Evaluation of carcinogenic hazard of diesel engine exhaust needs to consider revolutionary changes in diesel technology. Regulatory Toxicology and Pharmacology : RTP 63 (2), pp. 225-258. DOI: 10.1016/j.yrtph.2012.04.005.

McCormick, R. (2012): Low-Temperature Biodiesel Research Reveals Potential Key to Successful Blend Performance (Fact Sheet), NREL Highlights, Research & Development. NREL (National Renewable Energy Laboratory). Online: http://www.afdc.energy.gov/uploads/publication/OSTI_53881.pdf

McKone, T. a. o. (2015): California Dimethyl Ether Multimedia Evaluation. Tier I. Final Draft. Edited by California Environmental Protection Agency. The University of California, Davis; The University of

California, Berkeley. Online:
http://www.arb.ca.gov/fuels/multimedia/meetings/DMETierIReport_Feb2015.pdf

Md Ishak, N. A. I.; Raman, I. A.; Yarmo, M. A. et al. (2015): Ternary phase behavior of water microemulsified diesel-palm biodiesel. Front. Energy 9 (2), pp. 162-169. DOI: 10.1007/s11708-015-0355-9.

Melo, T. C. C. de; Machado, G. B.; Belchior, C. R. et al. (2012): Hydrous ethanol-gasoline blends – Combustion and emission investigations on a Flex-Fuel engine. Fuel 97, pp. 796-804. DOI: 10.1016/j.fuel.2012.03.018.

Melo, T. C. C. de; Machado, G. B.; Oliveira, E. J. de et al. (2011): Different Hydrous Ethanol-Gasoline Blends – FTIR Emissions of a Flex-Fuel Engine and Chemical Properties of the Fuels. In: SAE Brasil 2011 Congress and Exhibit, OCT. 04, 2011: SAE International, Warrendale, PA, United States (SAE Technical Paper Series).

Melo, Tadeu C. Cordeiro de; Machado, Guilherme B.; Belchior, Carlos R.P.; Colaço, Marcelo J.; Barros, José E.M.; Oliveira, Edimilson J. de; Oliveira, Daniel G. de (2012): Hydrous ethanol-gasoline blends – Combustion and emission investigations on a Flex-Fuel engine. In: Fuel 97, S. 796-804. DOI: 10.1016/j.fuel.2012.03.018

Michalke, B.; Fernsebner, K. (2014): New insights into manganese toxicity and speciation. Review. Journal of Trace Elements in Medicine and Biology 28, pp. 106-116. DOI: 10.1016/j.jtemb.2013.08.005.

Mol, J. (2004): Industrial applications of olefin metathesis. Journal of Molecular Catalysis A: Chemical 213 (1), pp. 39-45. DOI: 10.1016/j.molcata.2003.10.049.

Möller, V. P.; de Vaal, Philip L. (2014) An Investigation into Lubrication and Oxide Breakdown During Load-Carrying Capacity Testing, Tribology Transactions, 57:5, 890-898, DOI:10.1080/10402004.2014.921964.

Monirul, I. M.; Masjuki, H. H.; Kalam, M. A. et al. (2015): A comprehensive review on biodiesel cold flow properties and oxidation stability along with their improvement processes. RSC Adv 5 (105), pp. 86631-86655. DOI: 10.1039/C5RA09555G.

Montenegro, R. E. & Meier, M. A. R. (2012): Lowering the boiling point curve of biodiesel by crossmetathesis. European Journal of Lipid Science and Technology 114 (1), pp. 55-62.

Morales, A.; Marmesat, S.; Ruiz-Méndez, M. V. et al. (2014): Formation of oxidation products in edible vegetable oils analyzed as FAME derivatives by HPLC-UV-ELSD. Food Research International 62, pp. 1080-1086. DOI: 10.1016/j.foodres.2014.05.063.

Morganti, K.; Al-Abdullah, M. & Zubail, A. (2015): Fuel Economy and CO2 Emissions Benefits of Octane-on-Demand Combustion in Spark-Ignition Engines. In The Combustion Institute (Ed.): Proceedings of the Australian Combustion Symposium. With assistance of Y. Yang, N. Smith. Australian Combustion Symposium. Melbourne, December 7-9, 2015. The University of Melbourne, pp. 68-71.

Morgott, D. A. (2014): Factors and Trends Affecting the Identification of a Reliable Biomarker for Diesel Exhaust Exposure. Critical Reviews in Environmental Science and Technology 44 (16), pp. 1795-1864. DOI: 10.1080/10643389.2013.790748.

Mudgal, S.; Sonigo, P.; Toni, A. de et al. (2013): Development of a risk assessment for health and environment from the use of metallic additives and a test methodology for that purpose. Final Report. Edited by D. CLIMAG European Commission. Online: https://ec.europa.eu/clima/sites/clima/files/transport/fuel/docs/bio_report_en.pdf

Mühlbauer, W.; Zöllner, C.; Lehmann, S. et al. (2016): Correlations between physicochemical properties of emitted diesel particulate matter and its reactivity. Combustion and Flame 167, pp. 39-51. DOI: 10.1016/j.combustflame.2016.02.029.

Mueller-Langer, F.; Kaltschmitt, M. 2014. Biofuels from lignocellulosic biomass—a multi-criteria approach for comparing overall concepts. *Biomass Conversion and Biorefinery*. DOI: 10.1007/s13399-014-0125-7

Munack, A.; Pabst, C.; Fey, B. et al. (2012): Lowering of the boiling curve of biodiesel by metathesis. Final project report, UFOP project number 540/085. Braunschweig, Coburg and Karlsruhe. Online: http://www.ufop.de/files/7913/8063/2076/20120827_Abschlussb_final_eng_korr_Munack_Aend.pdf

Nadkarni, R. A. (2007): Guide to ASTM test methods for the analysis of petroleum products and lubricants. 2nd ed. West Conshohocken PA: ASTM International (ASTM's manual series, manual 44-2nd).

Namasivayam, A. M.; Korakianitis, T.; Crookes, R. J. et al. (2010): Biodiesel, emulsified biodiesel and dimethyl ether as pilot fuels for natural gas fuelled engines. Applied Energy 87 (3), pp. 769-778. DOI: 10.1016/j.apenergy.2009.09.014.

Naumann, K.; Oehmichen, K.; Remmele, E.; Thuneke, K.; Schröder, J.; Zeymer, M.; Zech, K.; Müller-Langer, F.. 2016. *Monitoring Biokraftstoffsektor*. DBFZ-Report Nr. 11. Leipzig. ISBN 978-3-946629-04-7 (based on F. O. LICHT (2016): World Ethanol & Biofuels Report Bd. 2008-2016).

Naumann K, Oehmichen K, Zeymer M, Müller-Langer F, Kröger M, Majer S. 2011. *Basisinformationen zur Entwicklung des Biokraftstoffsektors bis 2011*. Thrän, D.; Pfeifer, D. (Hrsg.): Schriftenreihe des BMU-Förderprogramms „Energetische Biomassenutzung". Band 3. Leipzig. ISSN 2192-1806.

Naziri, E.; Consonni, R. & Tsimidou, M. Z. (2014): Squalene oxidation products. Monitoring the formation, characterisation and pro-oxidant activity. Eur. J. Lipid Sci. Technol. 116 (10), pp. 1400-1411. DOI: 10.1002/ejlt.201300506.

Nickel, A.; Ung, T.; Mkrtumyan, G. et al. (2012): A Highly Efficient Olefin Metathesis Process for the Synthesis of Terminal Alkenes from Fatty Acid Esters. Top. Catal. 55, pp. 518-523. DOI: 10.1007/s11244-012-9830-2.

Niemi, S.; Vauhkonen, V.; Mannonen, S.; Ovaska, T.; Nilsson, O.; Sirviö, K. et al. (2016): Effects of wood-based renewable diesel fuel blends on the performance and emissions of a non-road diesel engine. In: Fuel 186, S. 1-10. DOI: 10.1016/j.fuel.2016.08.048.

Niven, R. K. (2005): Ethanol in gasoline: Environmental impacts and sustainability review article. Renewable and Sustainable Energy Reviews 9 (6), pp. 535-555. DOI: 10.1016/j.rser.2004.06.003.

Northrop, W. F.; Bohac, S. V.; Chin, J.-Y. et al. (2011): Comparison of Filter Smoke Number and Elemental Carbon Mass from Partially Premixed Low Temperature Combustion in a Direct-Injection Diesel Engine. J. Eng. Gas Turbines Power 133 (10), p. 102804. DOI: 10.1115/1.4002918.

NREL – National Renewable Energy Laboratory (2016): New Catalyst Reduces Wasted Carbon in Biofuel Process, Lowers Cost. Highlights in Research & Development, NREL/FS-5100-65850. Refers to Behl et al. (Energy & Fuels 2015) and Schaidle et al. (ACS Catalysis 2015). Online: http://www.nrel.gov/docs/fy16osti/65850.pdf

Ntziachristos, L.; Papadimitriou, G.; Ligterink, N. et al. (2016): Implications of diesel emissions control failures to emission factors and road transport NOx evolution. Atmospheric Environment 141, pp. 542-551. DOI: 10.1016/j.atmosenv.2016.07.036.

Nylund, N.-O.; Aakko, P.; Niemi, S. et al. (2005): Alcohols/ethers as oxygenates in diesel fuel: properties of blended fuels and evaluation of practical experiences. IEA AMF Annex XXVI Final Report. TEC TransEnergy Consulting Ltd, Befri Consult. IEA-AMF. Online: http://www.iea-amf.org/app/webroot/files/file/Annex%20Reports/AMF_Annex_26.pdf

NZGOV – New Zealand Government. 2012. *Vehicle Exhaust Emissions Amendment 2012*. Online: http://www.nzta.govt.nz/resources/rules/vehicle-exhaust-emissions-amendment-2012/ (09.12.2016)

NZGOV2016 – Ministry of Business, Innovation and Employment. 2016. *Duties, taxes and direct levies on motor fuels in New Zealand – From 1 July 2016*. Online: http://www.mbie.govt.nz/info-services/sectors-industries/energy/liquid-fuel-market/duties-taxes-and-direct-levies-on-motor-fuels-in-new-zealand (13.12.2016)

OECD – Organisation for Economic Co-operation and Development. 2016. *ISRAEL'S GREEN TAX ON CARS*. OECD Environment Policy Papers. 5. DOI: 10.1787/5jlv5rmnq9wg-en

Oestreich, D.; Lautenschütz, L.; Arnold, U. et al. (2017): Reaction kinetics and equilibrium parameters for the production of oxymethylene dimethyl ethers (OMEs) from methanol and formaldehyde. Chem. Eng. Sci. 163, 92–104. DOI: 10.1016/j.ces.2016.12.037.

Oestreich, D.; Lautenschütz, L.; Arnold, U. et al. (2018): Production of oxymethylene dimethyl ether (OME)-hydrocarbon fuel blends in a one-step synthesis/extraction procedure. Fuel 214, 39–44. DOI: 10.1016/j.fuel.2017.10.116.

Ogunkoya, D.; Fang, T. (2015): Engine performance, combustion, and emissions study of biomass to liquid fuel in a compression-ignition engine. In: Energy Conversion and Management 95, S. 342-351. DOI: 10.1016/j.enconman.2015.02.041.

Ohsedo, Y. (2016): Low-molecular-weight organogelators as functional materials for oil spill remediation. Polym. Adv. Technol. 27 (6), pp. 704-711. DOI: 10.1002/pat.3712.

Oliveira, M. B.; Pratas, M. J.; Queimada, A. J. et al. (2012): Another look at the water solubility in biodiesels: Further experimental measurements and prediction with the CPA EoS. Fuel 97, pp. 843-847. DOI: 10.1016/j.fuel.2012.03.022.

Onasch, T. B.; Trimborn, A.; Fortner, E. C. et al. (2012): Soot Particle Aerosol Mass Spectrometer. Development, Validation, and Initial Application. Aerosol Science and Technology 46 (7), pp. 804-817. DOI: 10.1080/02786826.2012.663948.

O'Neil G.W.; Culler A.R.; Williams, J. R. et al. (2015): Production of Jet Fuel Range Hydrocarbons as a Coproduct of Algal Biodiesel by Butenolysis of Long-Chain Alkenones. Energy Fuels 29, pp. 922-930. DOI: 10.1021/ef502617z.

Ooi, J. B.; Ismail, H. M.; Swamy, V. et al. (2016): Graphite Oxide Nanoparticle as a Diesel Fuel Additive for Cleaner Emissions and Lower Fuel Consumption. Energy Fuels. DOI: 10.1021/acs.energyfuels.5b02162.

Østerstrøm, F. F.; Anderson, J. E.; Mueller, S. A. et al. (2016): Oxidation Stability of Rapeseed Biodiesel/Petroleum Diesel Blends. Energy Fuels 30 (1), pp. 344-351. DOI: 10.1021/acs.energyfuels.5b01927.

Ouda, M.; Yarce, G.; White, R. J. et al. (2017): Poly(oxymethylene) dimethyl ether synthesis – a combined chemical equilibrium investigation towards an increasingly efficient and potentially sustainable synthetic route. React. Chem. Eng. DOI: 10.1039/C6RE00145A.

Oyedun, A. O.; Kumar, A.; Oestreich, D. et al. (2018): The development of the production cost of oxymethylene ethers as diesel additives from biomass. Biofuels, Bioprod. Biorefin.. DOI: 10.1002/bbb.1887.

Pabst, C. (2014): Wechselwirkungen von Kraftstoffgemischen mit hohem Biogenitätsgehalt am Beispiel eines Motors mit SCR-Abgasnachbehandlung. Dissertation (in German). Technische Universität Carolo-Wilhelmina zu Braunschweig, Braunschweig.

Palash, S. M.; Kalam, M. A.; Masjuki, H. H. et al. (2013): Impacts of biodiesel combustion on NOx emissions and their reduction approaches. Renewable and Sustainable Energy Reviews 23, pp. 473-490. DOI: 10.1016/j.rser.2013.03.003.

Park, S.; Choi, B. & Oh, B.-S. (2011): A combined system of dimethyl ether (DME) steam reforming and lean NOx trap catalysts to improve NOx reduction in DME engines. International Journal of Hydrogen Energy 36 (11), pp. 6422-6432. DOI: 10.1016/j.ijhydene.2011.02.124.

Park, S. H., Lee, C. S. (2013): Combustion performance and emission reduction characteristics of automotive DME engine system. Progress in Energy and Combustion Science 39 (1), pp. 147-168. DOI: 10.1016/j.pecs.2012.10.002.

Park, S. H. & Lee, C. S. (2014): Applicability of dimethyl ether (DME) in a compression ignition engine as an alternative fuel. Energy Conversion and Management 86, pp. 848-863. DOI: 10.1016/j.enconman.2014.06.051.

Patel, A. R. & Dewettinck, K. (2015): Comparative evaluation of structured oil systems: Shellac oleogel, HPMC oleogel, and HIPE gel. European Journal of Lipid Science and Technology: EJLST 117 (11), pp. 1772-1781. DOI: 10.1002/ejlt.201400553.

Pavlovic, J.; Marotta, A. & Ciuffo, B. (2016): CO2 emissions and energy demands of vehicles tested under the NEDC and the new WLTP type approval test procedures. Applied Energy 177, pp. 661-670. DOI: 10.1016/j.apenergy.2016.05.110.

Pearson, R. J.; Turner, J. W.; Bell, A. et al. (2014): Iso-stoichiometric fuel blends: Characterisation of physicochemical properties for mixtures of gasoline, ethanol, methanol and water. Proceedings of the Institution of Mechanical Engineers, Part D: Journal of Automobile Engineering 229 (1), pp. 111-139. DOI: 10.1177/0954407014529424.

Pedata, P.; Stoeger, T.; Zimmermann, R. et al. (2015): "Are we forgetting the smallest, sub 10 nm combustion generated particles?". Particle and Fibre Toxicology 12, p. 34. DOI: 10.1186/s12989-015-0107-3.

Penconek, A. & Moskal, A. (2016): The influence of pH and concentration of mucins on diesel exhaust particles (DEPs) transport through artificial mucus. Journal of Aerosol Science 102, pp. 83-95. DOI: 10.1016/j.jaerosci.2016.09.001.

Pereira, T. C.; Conceicao, C. A. F.; Khan, A. et al. (2016): Application of electrochemical impedance spectroscopy: A phase behavior study of babassu biodiesel-based microemulsions. Spectrochimica Acta. Part A, Molecular and Biomolecular Spectroscopy 168, pp. 60-64. DOI: 10.1016/j.saa.2016.05.034.

Perimensis A., Majer S., Zech K., Holland M., Müller-Lange F. 2010. *Technology Opportunities and Strategies Towards Climate Friendly Transport (TOSCA)*. WP 4 Report. Lifecycle Assessment of Transportation Fuels.

Pillai, S. K.; Hamoudi, S. & Belkacemi, K. (2013): Functionalized value-added products via metathesis of methyloleate over methyltrioxorhenium supported on ZnCl2-promoted mesoporous alumina. Fuel 110, pp. 32-39. DOI: 10.1016/j.fuel.2012.10.040.

Pirjola, L.; Parviainen, H.; Lappi, M. et al. (2004): A Novel Mobile Laboratory for "Chasing" City Traffic. SAE Technical Paper 2004-01-1962. In: 2004 SAE Fuels & Lubricants Meeting & Exhibition, JUN. 08, 2004: SAE International, Warrendale, PA, United States (SAE Technical Paper Series).

Platt, S. M.; El Haddad, I.; Zardini, A. A. et al. (2013): Secondary organic aerosol formation from gasoline vehicle emissions in a new mobile environmental reaction chamber. Atmos. Chem. Phys. 13 (18), pp. 9141-9158. DOI: 10.5194/acp-13-9141-2013.

Poitras, M.-J.; Rosenblatt, D. & Goodman, J. (2015): Impact of Ethanol and Isobutanol Gasoline Blends on Emissions from a Closed-Loop Small Spark-Ignited Engine. In: SAE 2015 World Congress & Exhibition, APR. 21, 2015: SAE International, Warrendale, PA, United States (SAE Technical Paper Series). Online: http://papers.sae.org/2015-01-1732/

Polat, S. (2016): An experimental study on combustion, engine performance and exhaust emissions in a HCCI engine fuelled with diethyl ether-ethanol fuel blends. Fuel Processing Technology 143, pp. 140-150. DOI: 10.1016/j.fuproc.2015.11.021.

Popovicheva, O.; Engling, G.; Lin, K.-T. et al. (2015): Diesel/biofuel exhaust particles from modern internal combustion engines. Microstructure, composition, and hygroscopicity. Fuel 157, pp. 232-239. DOI: 10.1016/j.fuel.2015.04.073.

Popovicheva, O. B. (2014): Microstructure and Chemical Composition of Diesel and Biodiesel Particle Exhaust. Aerosol Air Qual. Res. DOI: 10.4209/aaqr.2013.11.0336.

Price, P.; Stone, R.; Collier, T. et al. (2006): Dynamic Particulate Measurements from a DISI Vehicle: A Comparison of DMS500, ELPI, CPC and PASS. In: SAE 2006 World Congress & Exhibition, APR. 03, 2006: SAE International, Warrendale, PA, United States (SAE Technical Paper Series).

Prokopowicz, A.; Zaciera, M.; Sobczak, A.; Bielaczyc, P.; Woodburn, J. (2015): The effects of neat biodiesel and biodiesel and HVO blends in diesel fuel on exhaust emissions from a light duty vehicle with a diesel engine. In: Environmental Science & Technology 49 (12), S. 7473-7482. DOI: 10.1021/acs.est.5b00648.

PubChem (2017) Database Methylal, online: https://pubchem.ncbi.nlm.nih.gov/compound/Dimethoxymethane

Pullen, J. & Saeed, K. (2012): An overview of biodiesel oxidation stability. Renewable and Sustainable Energy Reviews 16 (8), pp. 5924-5950. DOI: 10.1016/j.rser.2012.06.024.

Raikos, V., Vamvakas, S.S., Kapolos, J., Koliadima, A., Karaiskakis, G. (2011) Identification and characterization of microbial contaminants isolated from stored aviation fuels by DNA sequencing and restriction fragment length analysis of a PCR-amplified region of the 16S rRNA gene, Fuel 90, pp. 695-700. DOI: 10.1016/j.fuel.2010.09.030

Rajesh Kumar, B. & Saravanan, S. (2016): Use of higher alcohol biofuels in diesel engines_ A review. Renewable and Sustainable Energy Reviews 60, pp. 84-115. DOI: 10.1016/j.rser.2016.01.085.

Rankovic, N.; Bourhis, G.; Loos, M. et al. (2015): Understanding octane number evolution for enabling alternative low RON refinery streams and octane boosters as transportation fuels. Fuel 150, pp. 41-47. DOI: 10.1016/j.fuel.2015.02.005.

Rantanen L, Linnalla R. (2005) NExBTL – biodiesel fuel of the second generation. SAE technical paper 2005-01-3771

Rauch. 2016. *Production of FT*. Presentation Nationaler Workshop Biotreibstoffe (29 September 2016). Online: http://www.nwbt.at/app/webroot/files/file/2016-09-29%20Workshop%20Biotreibstoffe/07_NWBT_092016_Rauch.pdf

Reddy, S. (2016): Development of a Thermodynamics-Based Fundamental Model for Prediction of Gasoline/Ethanol Blend Properties and Vehicle Driveability. CRC Project No. CM-138-15-1. CRC. Online: https://crcao.org/reports/recentstudies2016/CM-138-15-1/CRC%20Project%20CM-138-15-1%20-%20Final%20Report%20Revised%2016May2016.pdf.

Reham, S. S.; Masjuki, H. H.; Kalam, M. A. et al. (2015): Study on stability, fuel properties, engine combustion, performance and emission characteristics of biofuel emulsion. Renewable and Sustainable Energy Reviews 52, pp. 1566-1579. DOI: 10.1016/j.rser.2015.08.013.

Reiter, A. M.; Wallek, T.; Pfennig, A. et al. (2015): Surrogate Generation and Evaluation for Diesel Fuel. Energy Fuels 29 (7), pp. 4181-4192. DOI: 10.1021/acs.energyfuels.5b00422.

RES 2016a – Banasiak, J. 2016. *Biofuel quota (Act on Sustainable Biofuels)*. Online: http://www.res-legal.eu/en/search-by-country/denmark/single/s/res-t/t/promotion/aid/biofuel-quota-act-on-sustainable-biofuels-1/lastp/96/ (07.12.2016)

RES 2016b – Banasiak, J. 2016. *Tax regulation mechanism*. Online: http://www.res-legal.eu/en/search-by-country/denmark/single/s/res-t/t/promotion/aid/tax-regulation-mechanism-6/lastp/96/ (07.12.2016)

RES 2016c – Siniloo, G. 2016. *Biofuel quota (Distribution obligation system)*. Online: http://www.res-legal.eu/en/search-by-country/finland/single/s/res-t/t/promotion/aid/biofuel-quota-distribution-obligation-system/lastp/127/ (07.12.2016)

RES 2016d – Siniloo, G. 2016. *Tax regulation mechanism (Excise duty on liquid fuels)*. Online: http://www.res-legal.eu/en/search-by-country/finland/single/s/res-t/t/promotion/aid/tax-regulation-mechanism-excise-duty-on-liquid-fuels/lastp/127/ (07.12.2016)

Ribeiro, N. M.; Pinto, A.; Quintella, C. M. et al. (2007): The Role of Additives for Diesel and Diesel Blended (Ethanol or Biodiesel) Fuels: A Review. Energy Fuels 21, pp. 2433-2445.

Ricardo – Ricardo-AEA Ltd. 2016. *The role of natural gas and biomethane in the transport sector*. Final Report. Online: https://www.transportenvironment.org/sites/te/files/publications/2016_02_TE_Natural_Gas_Bio methane_Study_FINAL.pdf

RIS – Rechtsinformationssystem Bundeskanzleramt. 2016. *Bundesgesetz, mit dem die Mineralölsteuer an das Gemeinschaftsrecht angepaßt wird (Mineralölsteuergesetz 1995)*. Online: https://www.ris.bka.gv.at/GeltendeFassung.wxe?Abfrage=Bundesnormen&Gesetzesnummer=10 004908 (07.12.2016)

Robinson, A. L. (2014): Linking Tailpipe to Ambient: Phase 1-3. CRC Project A-74/E-96 Final Report. accessible via https://crcao.org/publications/atmosphereImpacts/index.html. Edited by CRC. Online: https://crcao.org/reports/recentstudies2014/A-74-E-96%20Phases%201-3/CRC%20A74_E96%20Phase%201-3%20Final%20Report_May2014.pdf.

Rockstroh, T.; Floweday, G. & Wilken, C. (2016): Options for Use of GTL Naphtha as a Blending Component in Oxygenated Gasoline. SAE Int. J. Fuels Lubr. 9 (1), pp. 191-202. DOI: 10.4271/2016-01-0879.

Roy, A.; Sonntag, D.; Cook, R. et al. (2016): Effect of Ambient Temperature on Total Organic Gas Speciation Profiles from Light-Duty Gasoline Vehicle Exhaust. Environmental Science & Technology 50 (12), pp. 6565-6573. DOI: 10.1021/acs.est.6b01081.

SAE, 1996: J1667 Recommended Practice, 1996 Snap Accelerated Smoke Test Procedure for Heavy Duty Powered Vehicles. Online: https://www.arb.ca.gov/enf/hdvip/saej1667.pdf

Sagiri, S. S.; Singh, V. K.; Pal, K. et al. (2015): Stearic acid based oleogels: a study on the molecular, thermal and mechanical properties. Materials science & engineering. C, Materials for biological applications 48, pp. 688-699. DOI: 10.1016/j.msec.2014.12.018.

Sánchez, R.; Sánchez, C.; Lienemann, C.-P. et al. (2015): Metal and metalloid determination in biodiesel and bioethanol. J. Anal. At. Spectrom. 30 (1), pp. 64-101. DOI: 10.1039/C4JA00202D.

Sarathy, S. M.; Kukkadapu, G.; Mehl, M. et al. (2016): Compositional effects on the ignition of FACE gasolines. Combustion and Flame 169, pp. 171-193. DOI: 10.1016/j.combustflame.2016.04.010.

Sarkar S, Kumar A, Sultana A. 2011. *Biofuels and biochemicals production from forest biomass in Western Canada.* Energy. 36(10): 6251-6262. DOI: 10.1016/j.energy.2011.07.024

Sauer, J.; Arnold, U.; Lautenschütz, L.; Oestreich, D.; Haltenort, P. 2016. *Routes from Methanol to OMEs – State of the Art and Future Potentials.* Institute of Catalysis Research and Technology (IKFT). Online: http://www.methanolmsa.com/wp-content/uploads/2016/12/J%E2%80%9Drg-Sauer.pdf

Sayin, C.; Ilhan, M.; Canakci, M.; Gumus, M.(2009): Effect of injection timing on the exhaust emissions of a diesel engine using diesel-methanol blends. Renewable Energy 34 (5), pp. 1261-1269. DOI: 10.1016/j.renene.2008.10.010.

Schaper, K.; Munack, A. & Krahl, J.: Parametrierung der physikalisch-chemischen Eigenschaften von Biokraftstoffen der 1,5. Generation. Dissertation, Göttingen.

Schifter, I.; Diaz, L.; Gómez, J. P. et al. (2013): Combustion characterization in a single cylinder engine with mid-level hydrated ethanol-gasoline blended fuels. Fuel 103, pp. 292-298. DOI: 10.1016/j.fuel.2012.06.002.

Schmitz, N.; Burger, J.; Ströfer, E. et al. (2016): From methanol to the oxygenated diesel fuel poly(oxymethylene) dimethyl ether. An assessment of the production costs. Fuel 185, pp. 67-72. DOI: 10.1016/j.fuel.2016.07.085.

Schmitz, N.; Homberg, F.; Berje, J. et al. (2015): Chemical Equilibrium of the Synthesis of Poly(oxymethylene) Dimethyl Ethers from Formaldehyde and Methanol in Aqueous Solutions. Ind. Eng. Chem. Res. 54 (25), pp. 6409-6417. DOI: 10.1021/acs.iecr.5b01148.

Schober, S. & Mittelbach, M. (2004): The impact of antioxidants on biodiesel oxidation stability. Eur. J. Lipid Sci. Technol. 106 (6), pp. 382-389. DOI: 10.1002/ejlt.200400954.

Schober, S. & Mittelbach, M. (2005): Influence of diesel particulate filter additives on biodiesel quality. European Journal of Lipid Science and Technology 107 (4), pp. 268-271. DOI: 10.1002/ejlt.200401115.

Scholwin F, Grope J. 2017. Innovative Solutions for Biomethane Production in Europe, Institute for Biogas, Waste Management & Energy, Presentation Fuels of the Future, Berlin January 2017

Schröder, O.; Pabst, C.; Munack, A. et al. (2016): Lowering the Boiling Curve of Biodiesel by Metathesis. MTZ Worldw. 77, pp. 64-69. DOI: 10.1007/s38313-015-0091-x.

SCOEL/SUM/53 – Nitrogen Dioxide, The Scientific Committee on Occupational Exposure Limits (2014). EU. Online: http://ec.europa.eu/social/BlobServlet?docId=12431&langId=en

SCOEL/SUM/75 – 1,3-butadiene, The Scientific Committee on Occupational Exposure Limits (2007). EU. Online: http://ec.europa.eu/social/BlobServlet?docId=3855&langId=en

SCOEL/SUM/89 – Nitrogen Monoxide, The Scientific Committee on Occupational Exposure Limits (2014). EU. Online: http://ec.europa.eu/social/BlobServlet?docId=12432&langId=en

Semelsberger, T. A.; Borup, R. L. & Greene, H. L. (2006): Dimethyl ether (DME) as an alternative fuel. Journal of Power Sources 156 (2), pp. 497-511. DOI: 10.1016/j.jpowsour.2005.05.082.

Sharma, C. S.; Ramanathan, K. & Li, W. (2012): Post-processing of vehicle emission test data for use in exhaust after-treatment modelling and analysis. Proceedings of the Institution of Mechanical Engineers, Part D: Journal of Automobile Engineering 226 (6), pp. 840-854. DOI: 10.1177/0954407011427812.

Shenghua, L.; Cuty Clemente, E. R.; Hu, T.; Wei, Y. (2007): Study of spark ignition engine fueled with methanol/gasoline fuel blends. Applied Thermal Engineering 27 (11-12), pp. 1904-1910. DOI: 10.1016/j.applthermaleng.2006.12.024.

Shields, L. G.; Suess, D. T. & Prather, K. A. (2007): Determination of single particle mass spectral signatures from heavy-duty diesel vehicle emissions for PM2.5 source apportionment. Atmospheric Environment 41 (18), pp. 3841-3852. DOI: 10.1016/j.atmosenv.2007.01.025.

Shiotani, H. & Goto, S. (2007): Studies of Fuel Properties and Oxidation Stability of Biodiesel Fuel. In: 2007 Fuels and Emissions Conference, JAN. 23, 2007: SAE International, Warrendale, PA, United States (SAE Technical Paper Series).

Silalertruksa T Gheewala H S, Sagisaka M, et. al. 2013. *Life cycle GHG analysis of rice straw bio-DME production and application in Thailand.* Applied Energy. 112: 560-567. DOI: 10.1016/j.apenergy.2013.06.028

Sileghem, L.; Coppens, A.; Casier, B.; Vancoillie, J.; Verhelst, S. (2014) Performance and emissions of iso-stoichiometric ternary GEM blends on a production SI engine. Fuel. 117: 286-293. DOI: 10.1016/j.fuel.2013.09.043

Simon, C. & Dörksen, H. (2016): Emission Reduction Potential of Various Diesel-Water Emulsions. MTZ Worldw. 77, pp. 74-81. DOI: 10.1007/s38313-015-0098-3.

Singer, A., Schröder, O., Pabst, C., Munack, A., Bünger, J., Ruck, W., Krahl, J.: Aging studies of biodiesel and HVO and their testing as neat fuel and blends for exhaust emissions in heavy-duty engines and passenger cars. Fuel 153, 595-603 (2015). DOI: 10.1016/j.fuel.2015.03.050

Singer, P. & Rühe, J. (2014): On the mechanism of deposit formation during thermal oxidation of mineral diesel and diesel/biodiesel blends under accelerated conditions. Fuel 133, pp. 245-252. DOI: 10.1016/j.fuel.2014.04.041.

Skeen, S. A.; Manin, J. & Pickett, L. M. (2015): Simultaneous formaldehyde PLIF and high-speed schlieren imaging for ignition visualization in high-pressure spray flames. Proceedings of the Combustion Institute 35 (3), pp. 3167-3174. DOI: 10.1016/j.proci.2014.06.040.

Smith, P. C.; Ngothai, Y.; Nguyen, Q. D. et al. (2010): The addition of alkoxy side-chains to biodiesel and the impact on flow properties. Fuel 89 (11), pp. 3517-3522. DOI: 10.1016/j.fuel.2010.06.014.

Song, H.; Liu, C.; Li, F. et al. (2017): A comparative study of using diesel and PODEn as pilot fuels for natural gas dual-fuel combustion. Fuel 188, pp. 418-426. DOI: 10.1016/j.fuel.2016.10.051.

Sonntag, D. B.; Gao, H. O. & Holmén, B. A. (2013): Comparison of particle mass and number emissions from a diesel transit bus across temporal and spatial scales. Transportation Research Part D: Transport and Environment 25, pp. 146-154. DOI: 10.1016/j.trd.2013.09.005.

Spitzer, P.; Fisicaro, P.; Seitz, S. et al. (2009): pH and electrolytic conductivity as parameters to characterize bioethanol. Accred Qual Assur 14 (12), pp. 671-676. DOI: 10.1007/s00769-009-0565-0.

Stålhammar, Per (2015): SCANIA – ED95 development. mot2030 Workshop on March 17 2015. Lunds Tekniska Högskola. Lund. Online: http://www.lth.se/fileadmin/mot2030/filer/11. Stalhammar_-_Scania_ED95_development.pdf

Steiger, W.; Stolte, U.; Scholz, I. et al. (2008): The CCS combustion system from Volkswagen. MTZ Worldw 69 (3), pp. 4-9. DOI: 10.1007/BF03226891.

Stein, R. A.; Anderson, J. E. & Wallington, T. J. (2013): An Overview of the Effects of Ethanol-Gasoline Blends on SI Engine Performance, Fuel Efficiency, and Emissions. SAE Int. J. Engines 6 (1), pp. 470-487. DOI: 10.4271/2013-01-1635.

Steiner, S.; Bisig, C.; Petri-Fink, A. et al. (2016): Diesel exhaust: current knowledge of adverse effects and underlying cellular mechanisms. Archives of Toxicology 90 (7), pp. 1541-1553. DOI: 10.1007/s00204-016-1736-5.

Steinigeweg S., Paul W., Meyer F., Hubel J. 2015. *Perspektiven und Potentiale von Low-Emission-LNG im Nordwesten*. Online: http://www.lng-nordwest.de/files/lng_downloads/Allgemein/Abschlussbericht_LowEmissionLNG_Abgabeversion.pdf (20.11.2016)

Strömberg, N.; Saramat, A. & Eriksson, H. (2013): Biodiesel degradation rate after refueling. Fuel 105, pp. 301-305. DOI: 10.1016/j.fuel.2012.09.072.

Sun, W.; Wang, G.; Li, S. et al. (2017): Speciation and the laminar burning velocities of poly(oxymethylene) dimethyl ether 3 (POMDME3) flames. An experimental and modeling study. Proceedings of the Combustion Institute 36 (1), pp. 1269-1278. DOI: 10.1016/j.proci.2016.05.058.

Swain, Pravat K.; Das, L. M.; Naik, S. N. (2011): Biomass to liquid. A prospective challenge to research and development in 21st century. In: Renewable and Sustainable Energy Reviews 15 (9), S. 4917-4933. DOI: 10.1016/j.rser.2011.07.061.

Swick, D.; Jaques, A.; Walker, J. C. et al. (2014): Gasoline risk management: a compendium of regulations, standards, and industry practices. Regulatory Toxicology and Pharmacology 70 (2 Suppl), S80-92. DOI: 10.1016/j.yrtph.2014.06.022.

Szybist, J. P.; McLaughlin, S. & Iyer, S. (2014): Emissions and Performance Benchmarking of a Prototype Dimethyl Ether-Fueled Heavy-Duty Truck. Edited by U.S. DEPARTMENT OF ENERGY. ORNL / UT-BATTELLE, LLC.

Tang, M.; Li, Q.; Xiao, L. et al. (2012): Toxicity effects of short term diesel exhaust particles exposure to human small airway epithelial cells (SAECs) and human lung carcinoma epithelial cells (A549). Toxicology Letters 215 (3), pp. 181-192. DOI: 10.1016/j.toxlet.2012.10.016.

Teng, H.; McCandless, J. C. & Schneyer, J. B. (2004): Thermodynamic Properties of Dimethyl Ether – An Alternative Fuel for Compression-Ignition Engines. In: SAE 2004 World Congress & Exhibition, MAR. 08, 2004: SAE International, Warrendale, PA, United States (SAE Technical Paper Series).

Terech, P. & Weiss, R. G. (1997): Low Molecular Mass Gelators of Organic Liquids and the Properties of Their Gels. Chem. Rev. 97 (8), pp. 3133-3160. DOI: 10.1021/cr9700282.

Thangaraja, J. & Kannan, C. (2016): Effect of exhaust gas recirculation on advanced diesel combustion and alternate fuels – A review. Applied Energy 180, pp. 169-184. DOI: 10.1016/j.apenergy.2016.07.096.

Thewes M, Muether M, Brassat A, Pischinger S, Sehr A. 2011. *Analysis of the Effect of Bio-Fuels on the Combustion in a Downsized DI SI Engine*. SAE Paper 2011-01-1991.

Thomas, G.; Feng, B.; Veeraragavan, A. et al. (2014): Emissions from DME combustion in diesel engines and their implications on meeting future emission norms: A review. Fuel Processing Technology 119, pp. 286-304. DOI: 10.1016/j.fuproc.2013.10.018.

Thomas, J. F.; Huff, S. P. & West, B. H. (2012): Fuel Economy and Emissions of a Vehicle Equipped with an Aftermarket Flexible-Fuel Conversion Kit. ORNL/TM-2011/483. ORNL.

Toner, S. M.; Sodeman, D. A. & Prather, K. A. (2006): Single Particle Characterization of Ultrafine and Accumulation Mode Particles from Heavy Duty Diesel Vehicles Using Aerosol Time-of-Flight Mass Spectrometry. Environmental Science & Technology 40 (12), pp. 3912-3921. DOI: 10.1021/es051455x.

Truex, T. J.; Pierson, W. R.; McKee, D. E. et al. (1980): Effects of Barium Fuel Additive and Fuel Sulfur Level on Diesel Particulate Emissions. Environmental Science & Technology 14, pp. 1121-1124.

Tucker, J. (2016): Diesel Fuel Low Temperature Operability Guide. CRC Report No. 671. CRC. Online: https://crcao.org/reports/recentstudies2016/CRC%20671/CRC%20671.pdf

Tunå P, Hulteberg C. 2014. *Woody biomass-based transportation fuels – A comparative techno-economic study*. Fuel. 117 Part B: 1020-1026. DOI: 10.1016/j.fuel.2013.10.019

Turner, J.; Pearson, R. J.; Dekker, E. et al. (2012): Evolution of alcohol fuel blends towards a sustainable transport energy economy. MIT Energy Initiative Symposium: Prospects for Flexible- und Bi-Fuel Light Duty Vehicles. Cambridge MA, 19 April. Online: http://www.methanolfuels.org/wp-content/uploads/2013/05/Lotus-Migration-to-Alcohol-Energy-Economy-Paper-MIT-2012.pdf

Turner, J.; Pearson, R. J.; Dekker, E. et al. (2013): Extending the role of alcohols as transport fuels using iso-stoichiometric ternary blends of gasoline, ethanol and methanol. Applied Energy 102, pp. 72-86. DOI: 10.1016/j.apenergy.2012.07.044.

UNEP – United Nations Environment Programme. 2016. *Status of Fuel Quality and Emissions Standards in Asia-Pacific, Updated January 2016*. Online: http://www.unep.org/transport/New/PCFV/pdf/Maps_Matrices/AP/matrix/AP_Matrix_Jan2016.pdf (23.11.2016)

Uy, D.; Ford, M. A.; Jayne, D. T. et al. (2014): Characterization of gasoline soot and comparison to diesel soot. Morphology, chemistry, and wear. Tribology International 80, pp. 198-209. DOI: 10.1016/j.triboint.2014.06.009.

van der Gaag, P. 2012. *Analysing Challenges of Producing Bio-LNG and building the infra for it*. Presentation Gloabal Biomethane Congress (10 October 2012). Brussels. Online: http://european-biogas.eu/wp-content/uploads/files/2013/11/12-Peter-van-der-Gaag-Analysing-Challenges-of-Producing-LBM-That-Is-Fit-to-Use-in-the-Existing-LNG-Distribution-Networks.pdf

Velázquez, S.; Apolinário, S. M.; Melo, E. H.; Elmajian, P. H. B. (2009): Report on Experiences of Ethanol Buses and Fuel Stations in São Paulo. CENBIO- Brazilian Reference Center on Biomass. São Paulo.

Venu, H. & Madhavan, V. (2017): Influence of diethyl ether (DEE) addition in ethanol-biodiesel-diesel (EBD) and methanol-biodiesel-diesel (MBD) blends in a diesel engine. Fuel 189, pp. 377-390. DOI: 10.1016/j.fuel.2016.10.101.

Verbeek R. Verbeek M. 2015. *LNG for trucks and ships: fact analysis Review of pollutant and GHG emissionsFinal*. TNO Report. Online: http://www.nationaallngplatform.nl/wp-content/uploads/2016/04/TNO-report_LNG_fact_analysis.pdf (20.11.2016)

Villela, A. C. S. & Machado, G. B. (2012): Multifuel Engine Performance, Emissions and Combustion Using Anhydrous and Hydrous Ethanol. In: 21st SAE Brasil International Congress and Exhibition, OCT. 02, 2012: SAE International, Warrendale, PA, United States (SAE Technical Paper Series).

Virgilio, A.; Amais, R. S.; Schiavo, D. et al. (2015): Dilute-and-Shoot Procedure for Determination of As, Cr, P, Pb, Si, and V in Ethanol Fuel by Inductively Coupled Plasma Tandem Mass Spectrometry. Energy Fuels 29 (7), pp. 4339-4344. DOI: 10.1021/acs.energyfuels.5b00434.

Vogt, R.; Kirchner, U.; Scheer, V. et al. (2003): Identification of diesel exhaust particles at an Autobahn, urban and rural location using single-particle mass spectrometry. Journal of Aerosol Science 34 (3), pp. 319-337. DOI: 10.1016/S0021-8502(02)00179-9.

Vorfalt, T. (2010): Olefinmetathese – Synthese und Mechanismus von Ruthenium-NHC-Komplexen. Dissertation (in German). TU Darmstadt, Darmstadt. Online: tuprints.ulb.tu-darmstadt.de/2175/1/Vorfalt_Tim_Dissertation_genehmigt.pdf

Vouitsis, E.; Ntziachristos, L.; Pistikopoulos, P. et al. (2009): An investigation on the physical, chemical and ecotoxicological characteristics of particulate matter emitted from light-duty vehicles. Environmental Pollution (Barking, Essex: 1987) 157 (8-9), pp. 2320-2327. DOI: 10.1016/j.envpol.2009.03.028.

Vukeya, H. M. (2015): The Use of Model Compounds to Investigate the Influence of Fuel Composition on the Thermo-Oxidative Stability of FAME/Diesel Blends. University of Cape Town. Online: http://open.uct.ac.za/bitstream/handle/11427/14132/thesis_ebe_2015_vukeya_hm.pdf?sequence=1 ; or https://open.uct.ac.za/bitstream/item/15172/thesis_ebe_2015_vukeya_hm.pdf?sequence=1

Wang, Z.; Qiao, X.; Hou, J. et al. (2011): Combustion and emission characteristics of a diesel engine fuelled with biodiesel/dimethyl ether blends. Proceedings of the Institution of Mechanical Engineers, Part D: Journal of Automobile Engineering 225 (12), pp. 1683-1691. DOI: 10.1177/0954407011406804.

Wang, D.; Niu, J.; Wang, Z. et al. (2015): Monoglyceride-based organogelator for broad-range oil uptake with high capacity. Langmuir: the ACS journal of surfaces and colloids 31 (5), pp. 1670-1674. DOI: 10.1021/acs.langmuir.5b00053.

Wang, Xiaochen; Chen, Zhenbin; Ni, Jimin; Liu, Saiwu; Zhou, Haijie (2015): The effects of hydrous ethanol gasoline on combustion and emission characteristics of a port injection gasoline engine. In: *Case Studies in Thermal Engineering* 6, S. 147-154. DOI: 10.1016/j.csite.2015.09.007

Wang, Z.; Liu, H.; Ma, X. et al. (2016): Homogeneous charge compression ignition (HCCI) combustion of polyoxymethylene dimethyl ethers (PODE). Fuel 183, pp. 206-213. DOI: 10.1016/j.fuel.2016.06.033.

Wang, Z.; Liu, H.; Zhang, J.; Wang, J.; Shuai, S. (2015): Performance, Combustion and Emission Characteristics of a Diesel Engine Fueled with Polyoxymethylene Dimethyl Ethers (PODE3-4)/ Diesel Blends. In: Energy Procedia 75, S. 2337-2344. DOI: 10.1016/j.egypro.2015.07.479.

Waynick, J. A. (2005): Characterization of Biodiesel Oxidation and Oxidation Products. CRC Project No. AVFL-2b. SwRI® Project No. 08-10721. Online: http://biodiesel.org/reports/20050801_gen-366.pdf

Weaver, J. W.; Exum, L. R. & Prieto, L. M. (2010): Gasoline Composition Regulations Affecting LUST Sites. EPA/600/R-10/001. Edited by U.S. EPA. Online: http://www.epa.gov/athens/publications

Weiss, M.; Bonnel, P.; Kühlwein, J. et al. (2012): Will Euro 6 reduce the NOx emissions of new diesel cars? – Insights from on-road tests with Portable Emissions Measurement Systems (PEMS). Atmospheric Environment 62, pp. 657-665. DOI: 10.1016/j.atmosenv.2012.08.056.

Westbrook, C. K. (2013): Biofuels Combustion. Annu. Rev. Phys. Chem. 64, pp. 201-219.

Westbrook, S. R. (2005): Evaluation and Comparison of Test Methods to Measure the Oxidation Stability of Neat Biodiesel. Online: http://biodiesel.org/reports/20050501_gen-362.pdf

Williams, A.; McCormick, R.; Lance, M. et al. (2014): Effect of Accelerated Aging Rate on the Capture of Fuel-Borne Metal Impurities by Emissions Control Devices. SAE Int. J. Fuels Lubr. 7 (2), pp. 471-479. DOI: 10.4271/2014-01-1500.

Wu, Y.-p. G. & Lin, Y.-f. (2012): Trace species formation pathway analysis of biodiesel engine exhaust. Applied Energy 91 (1), pp. 29-35. DOI: 10.1016/j.apenergy.2011.09.001.

Xu, M.; Hung, D.; Yang, J. et al. (2015): Flash-boiling Spray Behavior and Combustion in a Direct Injection Gasoline Engine. In: The Combustion Institute (Ed.): Proceedings of the Australian Combustion Symposium. With assistance of Y. Yang, N. Smith. Australian Combustion Symposium. Melbourne, December 7-9, 2015. The University of Melbourne, pp. 14-23.

Yaakob, Z.; Narayanan, B. N.; Padikkaparambil, S. et al. (2014): A review on the oxidation stability of biodiesel. Renewable and Sustainable Energy Reviews 35, pp. 136-153. DOI: 10.1016/j.rser.2014.03.055.

Yang, H.-H.; Lo, M.-Y.; Chi-Wei Lan, J. et al. (2007): Characteristics of trans,trans-2,4-decadienal and polycyclic aromatic hydrocarbons in exhaust of diesel engine fueled with biodiesel. Atmospheric Environment 41 (16), pp. 3373-3380. DOI: 10.1016/j.atmosenv.2006.12.028.

Yang, L.; Franco, V.; Mock, P. et al. (2015): Experimental Assessment of NOx Emissions from 73 Euro 6 Diesel Passenger Cars. Environmental Science & Technology 49 (24), pp. 14409-14415. DOI: 10.1021/acs.est.5b04242.

Yang, Z.; Hollebone, B. P.; Wang, Z. et al. (2014): Storage stability of commercially available biodiesels and their blends under different storage conditions. Fuel 115, pp. 366-377. DOI: 10.1016/j.fuel.2013.07.039.

Yang, Z.; Hollebone, B. P.; Wang, Z. et al. (2015): A preliminary study for the photolysis behavior of biodiesel and its blends with petroleum oil in simulated freshwater. Fuel 139, pp. 248-256. DOI: 10.1016/j.fuel.2014.08.061.

Yanowitz, J.; Christensen, E. & McCormick, R. L. (2011): Utilization of Renewable Oxygenates as Gasoline Blending Components. Technical Report NREL/TP-5400-50791. NREL – National Renewable Energy Laboratory. Online: http://www.nrel.gov/docs/fy11osti/50791.pdf

Yanowitz, J. & McCormick, R. L. (2016): Review: Fuel Volatility Standards and Spark-Ignition Vehicle Driveability. SAE Int. J. Fuels Lubr. 9 (2). DOI: 10.4271/2016-01-9072.

Yanowitz, J.; Ratcliff, M.; McCormick, R. et al. (2014): Compendium of Experimental Cetane Numbers. Technical Report NREL/TP-5400-61693. NREL – National Renewable Energy Laboratory.

Ying, W.; Genbao, L.; Wei, Z. et al. (2008): Study on the application of DME/diesel blends in a diesel engine. Fuel Processing Technology 89 (12), pp. 1272-1280. DOI: 10.1016/j.fuproc.2008.05.023.

Yinhui, W.; Rong, Z.; Yanhong, Q. et al. (2016): The impact of fuel compositions on the particulate emissions of direct injection gasoline engine. Fuel 166, pp. 543-552. DOI: 10.1016/j.fuel.2015.11.019.

Youn, I. M.; Park, S. H.; Roh, H. G. et al. (2011): Investigation on the fuel spray and emission reduction characteristics for dimethyl ether (DME) fueled multi-cylinder diesel engine with common-rail injection system. Fuel Processing Technology 92 (7), pp. 1280-1287. DOI: 10.1016/j.fuproc.2011.01.018.

Yusri, I. M.; Mamat, R.; Najafi, G.; Razman, A.; Awad, Omar I.; Azmi, W. H. et al. (2017): Alcohol based automotive fuels from first four alcohol family in compression and spark ignition engine. A review on engine performance and exhaust emissions. Renewable and Sustainable Energy Reviews 77, pp. 169-181. DOI: 10.1016/j.rser.2017.03.080.

Zacharof, N.; Tietge, U.; Franco, V. et al. (2016): Type approval and real-world CO2 and NOx emissions from EU light commercial vehicles. Energy Policy 97, pp. 540-548. DOI: 10.1016/j.enpol.2016.08.002.

Zavala, M.; Herndon, S. C.; Wood, E. C. et al. (2009): Comparison of emissions from on-road sources using a mobile laboratory under various driving and operational sampling modes. Atmos. Chem. Phys. 9 (1), pp. 1-14. DOI: 10.5194/acp-9-1-2009.

Zech, K.; Meisel, K.; Brosowski, A.; Toft, L. V.; Müller-Langer, F. 2016. Environmental and economic assessment of the Inbicon lignocellulosic ethanol technology. *Applied Energy*. 171:347-356. DOI: 10.1016/j.apenergy.2016.03.057

Zeman N. 2013. Biogas streams produce clean fuels. Renewable Energy Focus. 14(5): 24-26: DOI: 10.1016/S1755-0084(13)70093-0

Zhang, J.-J.; Huang, Z.; Wu, J.-H. et al. (2008): Combustion and performance of heavy-duty diesel engines fuelled with dimethyl ether. Proceedings of the Institution of Mechanical Engineers, Part D: Journal of Automobile Engineering 222 (9), pp. 1691-1703. DOI: 10.1243/09544070JAUTO783.

Zhang, T.; Nilsson, L. J.; Björkholtz, C. et al. (2016): Effect of using butanol and octanol isomers on engine performance of steady state and cold start ability in different types of Diesel engines. Fuel 184, pp. 708-717. DOI: 10.1016/j.fuel.2016.07.046.

Zhang X, Kumar A, Arnold U, Sauer J. 2014. *Biomass-derived Oxymethylene Ethers as Diesel Additives: A Thermodynamic Analysis*. Energy Procedia. 61: 1921-1924.
DOI: 10.1016/j.egypro.2014.12.242

Zhang, Y.; Voice, A.; Tzanetakis, T. et al. (2016): An Evaluation of Combustion and Emissions Performance with Low Cetane Naphtha Fuels in a Multicylinder Heavy-Duty Diesel Engine. J. Eng. Gas Turbines Power 138 (10), p. 102805.

Zhang, X., Oyedun, A.O., Kumar, A., Oestreich, D., Arnold, U., Sauer, J. (2016) An optimized process design for oxymethylene ether production from woody biomass-derived syngas, Biomass and Bioenergy 90, 7-14

Zhang, Z. H.; Cheung, C. S.; Chan, T. L. et al. (2011): Experimental investigation on regulated and unregulated emissions of a diesel/methanol compound combustion engine with and without diesel oxidation catalyst. Science of the Total Environment 408, pp. 865-872.

Zhang, Z. H.; Tsang, K. S.; Cheung, C. S.; Chan, T. L.; Yao, C. D. (2011): Effect of fumigation methanol and ethanol on the gaseous and particulate emissions of a direct-injection diesel engine. Atmospheric Environment 45 (11), pp. 2001-2008. DOI: 10.1016/j.atmosenv.2010.12.019.

Zhao, H.; Ge, Y.; Hao, C. et al. (2010): *Carbonyl compound emissions from passenger cars fueled with methanol/gasoline blends*. The Science of the total environment 408 (17), pp. 3607-3613. DOI: 10.1016/j.scitotenv.2010.03.046.

Zhao, H.; Ge, Y.; Tan, J. et al. (2011): Effects of different mixing ratios on emissions from passenger cars fueled with methanol/gasoline blends. Journal of Environmental Sciences 23 (11), pp. 1831-1838. DOI: 10.1016/S1001-0742(10)60626-2.

Zhao, Q.; Wang, H.; Qin, Z.-f. et al. (2011): Synthesis of polyoxymethylene dimethyl ethers from methanol and trioxymethylene with molecular sieves as catalysts. Journal of Fuel Chemistry and Technology 39 (12), pp. 918-923. DOI: 10.1016/S1872-5813(12)60003-6.

Zhao, X.; Ren, M. & Liu, Z. (2005): Critical solubility of dimethyl ether (DME)+diesel fuel and dimethyl carbonate (DMC)+diesel fuel. Fuel 84 (18), pp. 2380-2383. DOI: 10.1016/j.fuel.2005.05.014.

Zhao, Y.; Wang, Y.; Li, D.; Lei, X.; Liu, S. (2014) Combustion and emission characteristics of a DME (dimethyl ether)-diesel dual fuel premixed charge compression ignition engine with EGR (exhaust gas recirculation). Energy 72, pp. 608-617,
DOI: 10.1016/j.energy.2014.05.086

Zhen, X.; Wang, Y. (2015): An overview of methanol as an internal combustion engine fuel. Renewable and Sustainable Energy Reviews 52, pp. 477-493. DOI: 10.1016/j.rser.2015.07.083.

Zheng, Y.; Tang, Q.; Wang, T. et al. (2015a): Kinetics of synthesis of polyoxymethylene dimethyl ethers from paraformaldehyde and dimethoxymethane catalyzed by ion-exchange resin. Chemical Engineering Science 134, pp. 758-766. DOI: 10.1016/j.ces.2015.05.067.

Zheng, Y.; Tang, Q.; Wang, T. et al. (2015b): Molecular size distribution in synthesis of polyoxymethylene dimethyl ethers and process optimization using response surface methodology. Chemical Engineering Journal 278, pp. 183-189. DOI: 10.1016/j.cej.2014.10.056.

Zheng, Z.; Li, C.; Liu, H. et al. (2015): Experimental study on diesel conventional and low temperature combustion by fueling four isomers of butanol. Fuel 141, pp. 109-119. DOI: 10.1016/j.fuel.2014.10.053.

Zhu, R.; Hu, J.; Bao, X. et al. (2016): Tailpipe emissions from gasoline direct injection (GDI) and port fuel injection (PFI) vehicles at both low and high ambient temperatures. Environmental Pollution (Barking, Essex: 1987) 216, pp. 223-234. DOI: 10.1016/j.envpol.2016.05.066.

Zigler, B. (2012): Fuels for Advanced Combustion Engines. NREL, 2012. Online: https://energy.gov/sites/prod/files/2014/03/f10/ft002_zigler_2012_o.pdf

Zimmer, A.; Cazarolli, J.; Teixeira, R. M. et al. (2013): Monitoring of efficacy of antimicrobial products during 60days storage simulation of diesel (B0), biodiesel (B100) and blends (B7 and B10). Fuel 112, pp. 153-162. DOI: 10.1016/j.fuel.2013.04.062.

Zimmerman, N.; Pollitt, K.; Jeong, C.-H. et al. (2014): Comparison of three nanoparticle sizing instruments: The influence of particle morphology. Atmospheric Environment 86, pp. 140-147. DOI: 10.1016/j.atmosenv.2013.12.023.

www.ingramcontent.com/pod-product-compliance
Lightning Source LLC
Chambersburg PA
CBHW070735220326
41598CB00024BA/3435